REVISE PEARSON EDEXCEL A LEVEL
Mathematics
PRACTICE PAPERS Plus⁺

Series consultant: Harry Smith

Authors: Glyn Payne and Richard Carter

Also available to support your revision:

Revise A Level Revision Planner   9781292191546

The **Revise A Level Revision Planner** helps you to plan and organise your time, step-by-step, throughout your A Level revision. Use this book and wall chart to mastermind your revision.

Published by Pearson Education Limited, 80 Strand, London, WC2R 0RL.

www.pearsonschoolsandfecolleges.co.uk

Copies of official specifications for all Pearson qualifications may be found on the website:
qualifications.pearson.com

Text and illustrations © Pearson Education Ltd 2019
Typeset and illustrated by Tech-Set Ltd, Gateshead
Produced by Project One Publishing Solutions
Cover illustration by Miriam Sturdee

The rights of Glyn Payne and Richard Carter to be identified as authors of this work have been asserted
by them in accordance with the Copyright, Designs and Patents Act 1988.

First published 2019

22 21 20 19

10 9 8 7 6 5 4 3 2 1

British Library Cataloguing in Publication Data
A catalogue record for this book is available from the British Library

ISBN 978 1 292 21326 2

Printed in Slovakia by Neografia

Notes from the publisher

1. While the publishers have made every attempt to ensure that advice on the qualification and its
   assessment is accurate, the official specification and associated assessment guidance materials are the
   only authoritative source of information and should always be referred to for definitive guidance.

   Pearson examiners have not contributed to any sections in this resource relevant to examination
   papers for which they have responsibility.

2. Pearson has robust editorial processes, including answer and fact checks, to ensure the accuracy
   of the content in this publication, and every effort is made to ensure this publication is free of
   errors. We are, however, only human, and occasionally errors do occur. Pearson is not liable for
   any misunderstandings that arise as a result of errors in this publication, but it is our priority to
   ensure that the content is accurate. If you spot an error, please do contact us at
   resourcescorrections@pearson.com so we can make sure it is corrected.

Contents

Using this book

This book has been created to help you prepare for your exam by familiarising yourself with the approach of the papers and the exam-style questions. Unlike the exam, however, all of the questions have targeted hints, guidance and support in the margin to help you understand how to tackle them.

All questions also have fully worked solutions shown in the back of the book for you to refer to.

You may want to work through the papers at your own pace, to reinforce your knowledge of the topics and practise the skills you have gained throughout your course. Alternatively, you might want to practise completing a paper as if in an exam. If you do this, bear these points in mind:

- Use black ink or ball-point pen.
- Answer all questions and make sure your answers to parts of questions are clearly labelled.
- Answer the questions in the spaces provided – there may be more space than you need.
- In a real exam, you must show all your working out in order to get full credit.
- You may use a calculator in all three papers, but it must not have functions for algebra, differentiation and integration, or have retrievable formulae stored in them.
- Diagrams are not accurately drawn, unless otherwise indicated in the question.
- The marks for each question are shown in brackets. Use this as a guide to how much time to spend on each question.

Paper 1: Pure Mathematics 1

- The total number of marks available for each Paper 1: Pure Mathematics 1 is 100.
- You have 2 hours to complete Paper 1.

Paper 2: Pure Mathematics 2

- The total number of marks available for each Paper 2: Pure Mathematics 2 is 100.
- You have 2 hours to complete Paper 2.

Paper 3: Statistics and Mechanics

- The total number of marks available for each Paper 3: Statistics and Mechanics is 100.
- You have 2 hours to complete Paper 3.

Paper 1: Pure Mathematics 1

Answer all questions. Write your answers in the spaces provided.

1   Find the set of values for which

$$f(x) = x^3 - 5x^2 + 3x + 4$$

is a decreasing function.

(4)

....

....

....

....

....

....

....

....

....

....

....

....

....

....

....

....

....

(Total for Question 1 is 4 marks)

Revision Guide
page 39

Hint

Start by differentiating.

LEARN IT!

$f(x)$ is decreasing on an interval $[a, b]$ if $f'(x) \leqslant 0$ for all values of x in that interval.

Hint

You can give your answers as an inequality or using set notation.

Revision Guide
pages 58, 60
62

Hint Q2a

Evaluate f(4) then use
the answer as your
input for f.

Hint Q2b

Use partial fraction
techniques and
compare coefficients.

Hint Q2c

Consider the values
of f(x) when x = 2 and
when x → ∞. You can
use your answer to
part (b) to work out the
value of f(x) as x → ∞.

Hint Q2d

The domain of the
inverse function is the
same as the range of
the original function.

2 $f(x) = \dfrac{5x + 2}{2x - 1}, \quad x \geqslant 2$

(a) Find ff(4)

(2)

(b) Show that $f(x)$ can be written in the form $A + \dfrac{B}{2x - 1}$, where

   A and B are constants to be found.

(2)

(c) Hence, or otherwise, state the range of f.

(1)

(d) Find $f^{-1}(x)$, stating its domain.

(3)

$f(x) = \dfrac{5x + 2}{2x - 1}, \quad x \geqslant 2$

(Total for Question 2 is 8 marks)

3 The diagram shows the graph of

$$y = 24x - 8x^{\frac{3}{2}}$$

Revision Guide
page 45

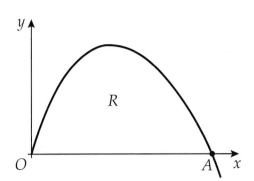

Hint Q3a

Solve $24x - 8x^{\frac{3}{2}} = 0$

Hint Q3b

Area $= \displaystyle\int_{0}^{a} (24x - 8x^{\frac{3}{2}})\, dx$

The graph crosses the x-axis at A.

(a) Find the value of the x-coordinate at A.

(2)

(b) Find the area, R, bounded by the curve and the x-axis between the origin, O, and point A.

(4)

(Total for Question 3 is 6 marks)

Revision Guide
pages 39, 40, 96

Hint Q4b

The stationary points

occur when $\dfrac{dy}{dx} = 0$

Hint Q4c

At a point of inflexion,

$\dfrac{d^2y}{dx^2} = 0$

Problem solving

$\dfrac{d^2y}{dx^2} = 0$ doesn't

guarantee a point of inflexion. You also need

the sign of $\dfrac{d^2y}{d^2x}$ to

change on either side of that point.

---

4 A curve, C, has equation

$$y = x^3 + 6x^2 + 9x + 5$$

(a) Find $\dfrac{dy}{dx}$ and $\dfrac{d^2y}{dx^2}$

(3)

(b) Verify that C has a stationary point when $x = -3$.

(2)

(c) Determine the nature of this stationary point, giving a reason for your answer.

(2)

(d) Find the coordinates of the point of inflexion on C.

(2)

(e) State, with a reason, whether C is concave or convex in the interval $[0, 1]$

(1)

(Total for Question 4 is 10 marks)

5 A circle has centre, P, at $(-2, 7)$ and radius of 5. The tangent to the circle at T passes through the point Q $(13, 12)$. The straight line PQ cuts the circle at W.

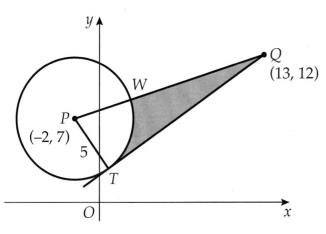

(a) Find the size of angle TPW in radians, giving your answer to 3 decimal places.

(4)

(b) Find the area of the shaded region TWQ, giving your answer to 2 decimal places.

(5)

Revision Guide
pages 76, 77

Hint Q5a

Make sure your calculator is in radians mode.

LEARN IT!

Area of sector $= \dfrac{1}{2} r^2 \theta$

Area of triangle
$= \dfrac{1}{2} ab \sin C$

Hint Q5b

Don't round any intermediate values. Use the memory functions on your calculator and write down at least 4 decimal places in your working.

(Total for Question 5 is 9 marks)

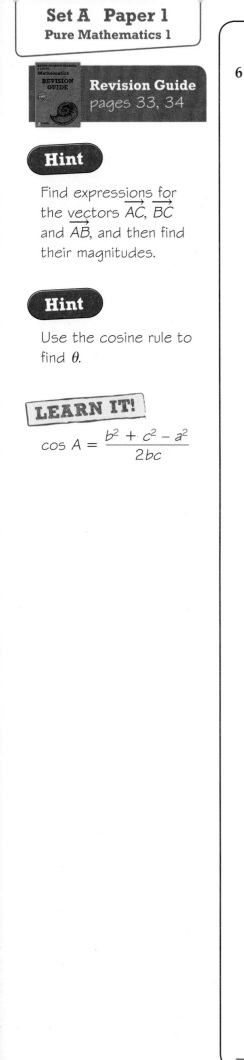

Revision Guide
pages 33, 34

Hint

Find expressions for the vectors $\overrightarrow{AC}$, $\overrightarrow{BC}$ and $\overrightarrow{AB}$, and then find their magnitudes.

Hint

Use the cosine rule to find θ.

LEARN IT!

$$\cos A = \frac{b^2 + c^2 - a^2}{2bc}$$

6

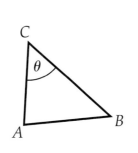

The diagram shows a sketch of triangle ABC. Given that, with reference to a fixed origin, O,

$$\overrightarrow{OA} = 2\mathbf{i} + \mathbf{j} - 3\mathbf{k}$$

$$\overrightarrow{OB} = 3\mathbf{i} - 2\mathbf{j} + \mathbf{k}$$

$$\overrightarrow{OC} = -\mathbf{i} + 3\mathbf{j} - \mathbf{k}$$

find the size of angle ACB, marked θ on the diagram.

(6)

(Total for Question 6 is 6 marks)

7 $f(x) = x^3 + x + 8$

(a) Show that $f(x) = 0$ has a root α in the interval $[-2, -1]$

(2)

(b) Find $f'(x)$

(1)

(c) Taking $x_0 = -2$ as a first approximation, apply the Newton–Raphson method three times to obtain an approximate value of α. Give your answer to 3 decimal places.

(4)

Revision Guide
page 99

Hint Q7a

You need to show that f(x) **changes sign** in the given interval. Make sure you state this fact in your answer.

Hint Q7c

The formula for the Newton–Raphson method is given in the formulae booklet:

$$x_{n+1} = x_n - \frac{f(x_n)}{f'(x_n)}$$

Hint Q7c

You should find that x_2 and x_3 round to the same value for α when written to 3 d.p.

(Total for Question 7 is 7 marks)

Revision Guide
page 70

Hint Q8a

The ratio between the consecutive terms is constant.

Problem solving

Use the common ratio to form an equation in k, then rearrange it. Remember that k is a positive constant.

8   The first three terms of a geometric sequence are

$2k, (3k + 4), (9k + 7)$

respectively, where k is a positive constant.

(a) Show that $9k^2 - 10k - 16 = 0$

(3)

(b) Find the value of k.

(2)

(c) Find the sum of the first 15 terms of this sequence, correct to 1 decimal place.

(3)

(Total for Question 8 is 8 marks)

9   A curve has parametric equations

$x = \cos\theta + \sin\theta$ and $y = 2\cos\theta + \sin\theta$

Find the Cartesian equation of the curve.

(5)

Revision Guide
page 86

Hint

Solve the equations simultaneously to find expressions for $\sin\theta$ and $\cos\theta$ in terms of x and y.

Problem solving

You can sometimes use $\sin^2\theta + \cos^2\theta = 1$ to eliminate the parameter in parametric equations.

Hint

Your final answer can be given implicitly. You don't need to be able to write it in the form $y = \dots$

(Total for Question 9 is 5 marks)

Revision Guide
page 94

Hint Q10a

Differentiate every term with respect to x, then collect all terms involving $\dfrac{dy}{dx}$ on one side, and the remaining terms on the other side.

LEARN IT!

Use the product rule, or learn the rule for differentiating xy terms implicitly:

$$\frac{d}{dx}(xy) = x\frac{dy}{dx} + y$$

Problem solving

The tangent to C will be parallel to the y-axis, which means the gradient is infinite.

10 The curve C has equation

$$x^2 + xy = 2y^2 + 63$$

(a) Find $\dfrac{dy}{dx}$ in terms of x and y.

(5)

(b) A point P lies on C. The tangent to C at the point P is parallel to the y-axis.

Find the possible coordinates of P.

(5)

(Total for Question 10 is 10 marks)

11 The point P is the point on the curve

$$x = 3\cot\left(\frac{\pi}{2} - y\right)$$

with y-coordinate $\frac{\pi}{6}$

Find the equation of the normal to the curve at P.

(8)

Revision Guide
pages 38, 92

Hint

The normal to the curve at P is perpendicular to the tangent at P.

Hint

Use $y = \frac{\pi}{6}$ to find the x-coordinate of P.

Problem solving

You are given x in terms of y, so first find $\frac{dx}{dy}$, then invert to find $\frac{dy}{dx}$.

(Total for Question 11 is 8 marks)

Revision Guide
pages 78–80, 83

LEARN IT!

You need to know the definitions of sec, cosec and cot. They are not given in the formulae booklet.

Hint Q12a

Rewrite cosec $2A$ and cot $2A$ in terms of $\sin 2A$ and $\cos 2A$, Then use the double angle formulae.

Hint Q12b

Make sure you have found all the possible answers in the given range – there should be two.

12 (a) Prove that $\operatorname{cosec} 2A - \cot 2A \equiv \tan A$

(4)

(b) Hence, or otherwise, solve for $0° < \theta < 180°$,

$$\operatorname{cosec}(4\theta - 20°) - \cot(4\theta - 20°) = \sqrt{3}$$

(4)

(Total for Question 12 is 8 marks)

13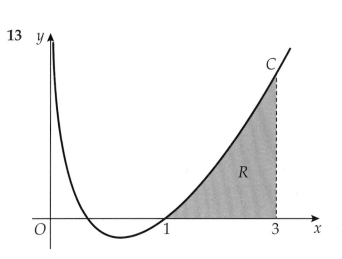

The diagram shows a sketch of part of the curve, C, with equation

$$y = (3x - 1)\ln x, \; x > 0$$

The finite region R is bounded by the curve C, the x-axis and the line with equation $x = 3$.

The table shows corresponding values of x and y, given to 4 decimal places, as appropriate.

x	1	1.5	2	2.5	3
y	0	1.4191	3.4657	5.9559	8.7889

(a) Use the trapezium rule, with all the values of y in the table, to obtain an estimate for the area of R, giving your answer to 3 decimal places.

(3)

(b) Explain how the trapezium rule could be used to obtain a more accurate estimate for the area of R.

(1)

(c) Show that the exact area of R can be written in the form $\frac{a}{b}\ln c + d$,

where a, b, c and d are integers to be found.

(7)

(In part (c), solutions based entirely on graphical or numerical methods are not acceptable.)

....

....

....

....

....

Revision Guide
pages 104, 108

Hint Q13a

There are five values given so you need to use the trapezium rule with four strips.

Hint Q13b

The question says 'explain' so give an answer in words.

Hint Q13c

Remember to set $\ln x$ as u when using integration by parts.

Problem solving

You will need to use integration by parts twice in part (c).

Watch out!

You need to give your answer to part (c) in the correct form. You could write out the values of a, b, c and d to make sure.

(Total for Question 13 is 11 marks)

TOTAL FOR PAPER IS 100 MARKS

Paper 2: Pure Mathematics 2

Answer all questions. Write your answers in the spaces provided.

1 The curve C has equation $y = f(x)$ where

$$f(x) = \frac{5x - 1}{x + 3}, \; x \neq -3$$

(a) Show that $f'(x) = \dfrac{16}{(x + 3)^2}$

(3)

The diagram shows the graph of $y = f(x)$. The point P with x-coordinate 1 lies on C.

The line L_1 is a tangent to C at P. The line L_2 is also a tangent to C at Q, and is parallel to L_1.

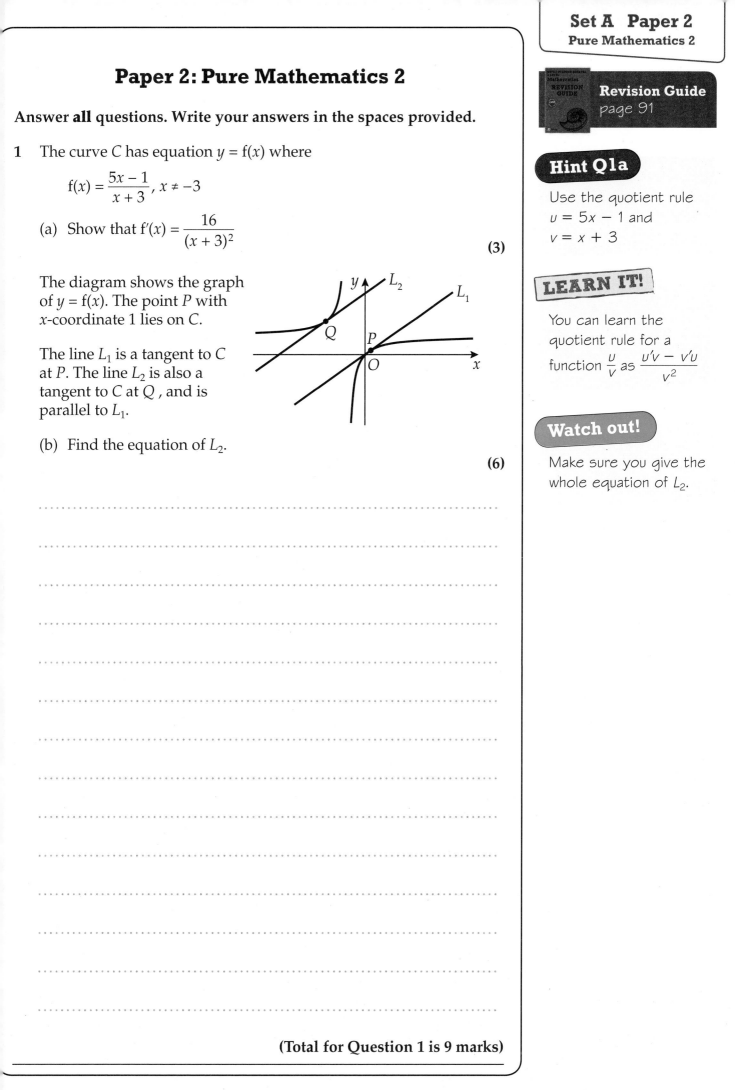

(b) Find the equation of L_2.

(6)

....

....

....

....

....

....

....

....

....

....

....

....

....

(Total for Question 1 is 9 marks)

Revision Guide
page 91

Hint Q1a

Use the quotient rule
$u = 5x - 1$ and
$v = x + 3$

LEARN IT!

You can learn the quotient rule for a function $\dfrac{u}{v}$ as $\dfrac{u'v - v'u}{v^2}$

Watch out!

Make sure you give the whole equation of L_2.

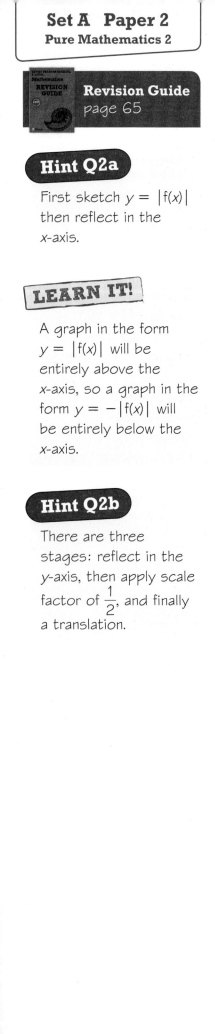

Revision Guide
page 65

Hint Q2a

First sketch $y = |f(x)|$ then reflect in the x-axis.

LEARN IT!

A graph in the form $y = |f(x)|$ will be entirely above the x-axis, so a graph in the form $y = -|f(x)|$ will be entirely below the x-axis.

Hint Q2b

There are three stages: reflect in the y-axis, then apply scale factor of $\frac{1}{2}$, and finally a translation.

2

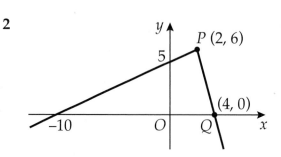

The diagram shows part of a graph with equation $y = f(x)$, $x \in \mathbb{R}$.

The graph consists of two line segments that meet at the point P (2, 6). The graph also crosses the axes at Q (4, 0) and two other points, (−10, 0) and (0, 5).

Sketch, on separate diagrams, the graphs of

(a) $y = -|f(x)|$

(2)

(b) $y = \frac{1}{2}f(-x) - 2$

(2)

Label the points P' and Q' on your sketch graphs.

(c) Does $f^{-1}(x)$ exist? Explain your answer.

(1)

....

....

....

....

....

....

....

....

....

....

....

....

(Total for Question 2 is 5 marks)

3 $g(x) = x^3 + ax^2 - ax + 80$, where a is a constant.

Given that $g(-8) = 0$

(a) (i) show that $a = 6$

   (ii) express $g(x)$ as a product of two algebraic factors.

   (4)

Given that $2\log_5(x + 3) + \log_5 x - \log_5(3x - 16) = 1$

(b) show that $x^3 + 6x^2 - 6x + 80 = 0$

   (4)

(c) Hence explain why

$$2\log_5(x + 3) + \log_5 x - \log_5(3x - 16) = 1$$

has no real roots.

   (2)

Revision Guide
pages 7, 23, 48, 49

Hint Q3a(i)

Substitute $x = -8$ in $g(x) = 0$, then solve to find a.

Hint Q3a(ii)

Since $g(-8) = 0$, $(x + 8)$ will be a factor of $g(x)$.

Hint Q3b

Use the laws of logarithms to simplify the given equation.

LEARN IT!

$a \log b = \log b^a$

$\log a + \log b = \log ab$

$\log a - \log b = \log \dfrac{a}{b}$

(Total for Question 3 is 10 marks)

4

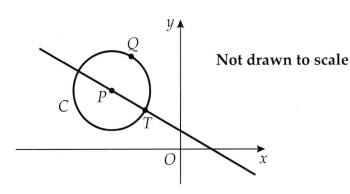

Not drawn to scale

The circle C has centre P.

Points Q and T lie on C.

The equation of the line PT is $2x + 3y = 6$

The line through P perpendicular to PT passes through the point Q $(-5, 14)$.

(a) Show that an equation of the line PQ is

$$3x - 2y = -43$$

(3)

(b) Find an equation for C.

(4)

The line with equation $3x - 2y = k$ is a tangent to C at the point T.

(c) Find the value of the constant k.

(3)

Hint Q4a

Rearrange the equation of PT into the form $y = mx + c$ to find the gradient.

Then use $\dfrac{-1}{m}$ to find the perpendicular gradient.

Hint Q4b

Find the coordinates of P and use

radius = distance PQ.

Hint Q4c

Work out the coordinates of T.

(Total for Question 4 is 10 marks)

5   The curve C has equation $y = 2x^3 - 3x$

The point $P\ (2, 10)$ lies on C.

Use differentiation from first principles to find the value of the gradient of the tangent to C at P.

(5)

Revision Guide
page 35

LEARN IT!

The rule for differentiating from first principles is given in the formulae booklet:

$$f'(x) = \lim_{h \to 0} \frac{f(x + h) - f(x)}{h}$$

Hint

Take care when expanding brackets and use correct mathematical language and notation.

(Total for Question 5 is 5 marks)

Revision Guide
page 84

Problem solving

Don't round any intermediate values. Write down at least 4 decimal places or use the memory functions on your calculator.

Hint Q6b

Check that all your final solutions are within the specified range.

Problem solving

You can use a sketch graph to check that you have found all the possible solutions.

6   (a)   Write $6 \cos \theta + 11 \sin \theta$ in the form

$$R \cos(\theta - \alpha) \text{ where } R \geqslant 0 \text{ and } 0 < \alpha < \frac{\pi}{2}$$

(3)

(b)   Hence, solve the equation

$$6 \cos \theta + 11 \sin \theta = 8$$

for $0 \leqslant \theta \leqslant 2\pi$, giving your answers to 2 decimal places.

(5)

(Total for Question 6 is 8 marks)

7   A hot metal sphere is cooled by placing it into a liquid. The
    temperature, $T\,°C$, after t minutes in the liquid, is given by

$$T = 340\,e^{-0.04t} + 23, \; t \geqslant 0$$

(a) Find the temperature of the sphere as it enters the liquid.

                                                                    (1)

Revision Guide
page 52

Hint Q7a

Substitute $t = 0$ into
the equation.

(b) Find the value of t when $T = 300$, giving your answer to
     3 significant figures.

                                                                    (2)

(c) Find the rate at which the temperature of the sphere is
     decreasing at the instant when $t = 45$. Give your answer
     in $°C$ per minute, to 3 significant figures.

                                                                    (3)

Hint Q7c

The rate of change of
the temperature
will be given by $\dfrac{dT}{dt}$.
This value will be
negative, because the
temperature is falling.

(d) Sketch a graph of T against t, showing clearly any points where
     the graph crosses or touches the axes, and any asymptotes.

                                                                    (2)

3 hours after the metal sphere was put into the liquid to cool, its
temperature was measured as $21.6\,°C$.

(e) Using this information, evaluate the model, explaining your
     reasoning.

                                                                    (1)

Watch out!

Sketch graphs should
still be drawn neatly.
Make sure you label
the axes, the origin, any
asymptotes, and any
points of intersection
with the axes.

Modelling

Compare the recorded
value with the value
predicted by the
model.

(Total for Question 7 is 9 marks)

Revision Guide
pages 39, 58

8 $\dfrac{29 - 6x - 2x^2}{(x + 4)(3 - x)} \equiv A + \dfrac{B}{x + 4} + \dfrac{C}{3 - x}$

(a) Find the values of the constants A, B and C.

(4)

$$f(x) = \dfrac{29 - 6x - 2x^2}{(x + 4)(3 - x)},\ x > 3$$

(b) Prove that $f(x)$ is a decreasing function.

(3)

Hint Q8a

One approach for this question is to multiply both sides by $(x + 4)(3 - x)$ then compare coefficients.

Hint Q8b

Differentiate the three terms on the right-hand side of the original identity. You need to show that the expression for $f'(x) \leq 0$ for all values of $x > 3$.

....

....

....

....

....

....

....

....

....

....

....

....

....

....

....

....

....

....

(Total for Question 8 is 7 marks)

Revision Guide
page 103

9 Using a suitable substitution, or otherwise, use an algebraic method to find the exact value of

$$\int_{\frac{\pi}{12}}^{\frac{\pi}{4}} \sin^4 2x \cos 2x \, dx$$

(6)

(Total for Question 9 is 6 marks)

10 Solve, for $-180° \leq \theta \leq 180°$

$$2 \sin \theta \tan \theta = \sin \theta + 6 \cos \theta$$

(9)

Revision Guide
page 79

Problem solving

When equations have terms in sin, cos and tan, it is usually better to express the equation in terms of sin and cos only.

Hint

Once you have written the equation in terms of sin and cos, you can factorise it.

(Total for Question 10 is 9 marks)

Revision Guide
page 75

Problem solving

Take out a factor of $9^{\frac{1}{2}}$ before using the binomial expansion.

Keep the $\left(\dfrac{4x}{9}\right)$ term in brackets so that you don't make any mistakes when squaring.

Watch out!

Remember to fully simplify all coefficients.

11 (a) Find the binomial expansion of

$$(9 + 4x)^{\frac{1}{2}}, \ |x| < \frac{9}{4}$$

in ascending powers of x, up to and including the term in x^2. Give each coefficient in its simplest form.

(5)

(b) Find the exact value of $(9 + 4x)^{\frac{1}{2}}$ when $x = 0.75$. Give your answer in the form $k\sqrt{3}$, where k is a constant to be found.

(2)

(c) Substitute $x = 0.75$ into your binomial expansion and hence find an approximate value for $\sqrt{3}$. Give your answer to 3 decimal places.

(3)

(Total for Question 11 is 10 marks)

12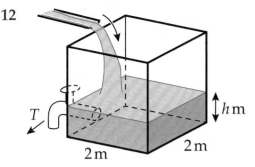

The diagram shows a water tank with a square base of side 2 m.
Water is flowing into the tank at a constant rate of $0.5\,\text{m}^3$ per
minute.

At time t minutes, the depth of the water in the tank is h metres.

When a tap, T, is opened at the bottom of the tank, water leaves the
tank at a rate of $0.4h\,\text{m}^3$ per minute.

(a) Show that t minutes after the tap has opened,

$$40\frac{\text{d}h}{\text{d}t} = 5 - 4h$$

(5)

When $t = 0, h = 0.6$

(b) Solve the differential equation from part (a) to show that

$$t = A\ln\left(\frac{B}{5 - 4h}\right)$$

where A and B are constants to be found.

(5)

(c) Find the value of t when $h = 0.9$. Give your answer to 2 decimal
places.

(2)

Revision Guide
pages 41, 97,
109

Modelling

Use the connection
between $\dfrac{\text{d}V}{\text{d}t}$ and $\dfrac{\text{d}h}{\text{d}t}$
to set up an equation
for $\dfrac{\text{d}h}{\text{d}t}$, the rate of
change of height with
respect to time.

Problem solving

Use separation of
variables to solve the
differential equation.
Remember to include a
constant of integration
and evaluate it using
the initial conditions
given.

(Total for Question 12 is 12 marks)

TOTAL FOR PAPER IS 100 MARKS

Paper 3: Statistics and Mechanics
SECTION A: STATISTICS

Revision Guide
pages 121, 145

Answer **all** questions. Write your answers in the spaces provided.

1 'Super Spicy' chilli powder is sold in packets. A shopkeeper measures the masses of the contents of a random sample of 80 packets of 'Super Spicy' chilli powder from his stock. The results are shown in the table.

Mass, w (g)	Midpoint, x (g)	Frequency, f
$391 \leqslant w < 395$	393	6
$395 \leqslant w < 398$	396.5	12
$398 \leqslant w < 402$	400	34
$402 \leqslant w < 407$	404.5	18
$407 \leqslant w < 410$	408.5	10

(You may use $\Sigma fx^2 = 12\,867\,128$)

A histogram is drawn and the class $395 \leqslant w < 398$ is represented by a rectangle of width 1.5 cm and height 6 cm.

(a) Calculate the width and height of the rectangle representing the class $402 \leqslant w < 407$

(3)

(b) Use linear interpolation to estimate the median mass of the contents of a packet of 'Super Spicy' chilli powder.

(2)

(c) Estimate the mean and the standard deviation of the mass of the contents of a packet of 'Super Spicy' chilli powder to 1 decimal place.

(3)

The shopkeeper claims that the mean mass of the contents of the packets is more than the stated mass. Given that the stated mass of a packet of 'Super Spicy' chilli powder is 400 g and that the actual standard deviation is 4 g:

(d) test, using a 2% level of significance, whether or not the shopkeeper's claim is justified. State your hypotheses clearly. (You may assume that the mass of the contents of a packet is normally distributed.)

(5)

(e) Using your answers to parts (b) and (c), comment on the assumption that the mass of the contents of a packet is normally distributed.

(1)

Hint Q1a

Work out the relationship between the area of the bar and the frequency.

Hint Q1d

The sample size is 80, so use a mean of 400 and a standard deviation of $\dfrac{4}{\sqrt{80}}$ when calculating your test statistic.

LEARN IT!

In a histogram, the area of each bar is proportional to the frequency.

Modelling

A normal distribution is symmetrical, so mean = median. This is one condition that can be used to determine whether a normal distribution is a suitable model.

(Total for Question 1 is 14 marks)

2 The table shows the mean daily temperature ($t°C$) and the mean daily rainfall (h mm) for the month of June 2015, for the seven places in the northern hemisphere from the large data set.

	A	B	C	D	E	F	G
$t°C$	12.8	13.3	16.8	15.1	13.8	24.7	26.4
h mm	1.52	0.81	0.58	1.10	1.67	1.60	8.60

(a) Calculate the product moment correlation coefficient for these data.

(1)

(b) Stating your hypotheses clearly, test, at the 5% level of significance, whether or not the product moment correlation coefficient is greater than 0.

(3)

(c) Using your knowledge of the large data set, suggest the names of the places labelled **F** and **G** in the table.

(1)

(d) Suggest how you could make better use of the large data set to investigate the relationship between the mean daily temperature and the mean daily rainfall, justifying your answer.

(2)

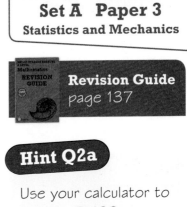

Revision Guide
page 137

Hint Q2a

Use your calculator to find the PMCC.

Hint Q2b

Write null and alternative hypotheses in terms of the population PMCC, p.

Hint Q2b

Compare your value of the product moment correlation coefficient with the 5% tabulated value for $n = 7$.

Remember to state your conclusion in the context of the original data.

(Total for Question 2 is 7 marks)

Revision Guide
pages 140, 142, 145

Hint Q3a

Use your calculator with the values given in the question. Write down your answer to four decimal places.

Hint Q3c

Use the inverse normal function on your calculator.

Hint Q3d

Remember to work out the variance for the sample, $\dfrac{\sigma^2}{n}$, before you use a hypothesis test.

3   A firm manufactures energy-saving halogen light bulbs. The random variable, X, represents the lifetime, in hours, of a particular light bulb.

X is normally distributed with mean 2000 hours and standard deviation 150 hours.

(a) Find $P(1900 < X < 2200)$

(1)

(b) The manufacturer claims that 95% of these light bulbs last longer than 1750 hours. Is this claim valid?

(2)

(c) Find the time, h hours, such that 99% of light bulbs last at least h hours.

(2)

The manufacturer makes a slight change to the process for producing the light bulbs. They now claim an increase in the mean lifetime of the light bulbs and a standard deviation of 120 hours.

(d) A sample of 30 light bulbs was chosen and found to have a mean lifetime of 2040 hours.

Stating your hypotheses clearly, and using a 1% level of significance, test whether or not these findings support the claim that the mean lifetime with the new process is more than 2000 hours.

(5)

(Total for Question 3 is 10 marks)

4 The Venn diagram shows the probabilities that students in Year 12 at a particular sixth form college study chemistry (C), maths (M) and history (H).

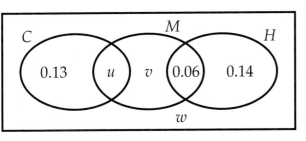

u, v and w are probabilities.

(a) If this Venn diagram represents the choices of 125 students, estimate how many take only chemistry or only history.

(2)

Given that the events M and H are independent, and that $P(M \mid C) = \dfrac{9}{22}$

(b) calculate the values of u, v and w.

(6)

Watch out!

The numbers on this Venn diagram can represent frequencies, elements or probabilities. If they are numbers less than 1, they probably represent probabilities.

Hint Q4b

Use the definitions of independent events and conditional probability to set up equations in u and v.

....

....

....

....

....

....

....

....

....

....

....

....

....

....

....

....

....

(Total for Question 4 is 8 marks)

Revision Guide
pages 131, 144

Problem solving

In part (a), you will need to use the binomial distribution twice: once for the seeds in a tray and once for the number of trays.

When using the normal distribution as an approximation for the binomial distribution, you will need to use a continuity correction. Read the question and choose the continuity correction carefully.

Modelling

Interpret your answer to part (c) in the context of the question. These answers can be quite subjective, so explain your conclusion as clearly and carefully as you can.

5   A seed company claims that 58% of its cucumber seeds germinate. A random selection of cucumber seeds is planted in 5 trays with 24 seeds in each tray.

   (a) Find the probability that in at least three of the trays, 16 or more seeds will germinate.

   (4)

   (b) State two conditions where the normal distribution can be used as an approximation to the binomial distribution.

   (1)

   A random sample of 120 cucumber seeds was planted and 75 of the seeds germinated.

   (c) Assuming a success rate of 58% germination, use a normal approximation to find the probability that at least 75 seeds will germinate.

   (4)

   (d) Does your answer to part (c) support the seed company's claim that 58% of its cucumber seeds germinate?

   (2)

(Total for Question 5 is 11 marks)

SECTION B: MECHANICS

Answer **all** questions. Write your answers in the spaces provided.

Unless otherwise indicated, whenever a numerical value of g is required, take $g = 9.8\,\mathrm{m\,s^{-2}}$ and give your answer to either 2 significant figures or 3 significant figures.

6   A particle is moving in a straight line. At time t seconds, $t \geq 0$, the acceleration of the particle, $a\ \mathrm{m\,s^{-2}}$, is given by

$$a = 2 \cos\left(\frac{\pi}{6}t\right)$$

(a) The velocity of the particle at $t = 0$ is $\dfrac{2}{\pi}\,\mathrm{m\,s^{-1}}$.

Find an expression for the velocity of the particle at t seconds.

(3)

(b) Find the velocity of the particle when $t = 5$.

(2)

Revision Guide
page 160

LEARN IT!

Integrate the expression for acceleration with respect to time to find an expression for velocity

Hint Q6a

Remember to include a constant of integration.

(Total for Question 6 is 5 marks)

Revision Guide
pages 160, 155

Hint Q7c

After differentiating to find the acceleration vector, you need to use **F** = *m***a**, which will give **F** as a vector. The question asks for the magnitude of **F**, so there is still work to do.

7   A particle, *P*, of mass 0.6 kg moves under the action of a single force **F** newtons.

At time *t* seconds, the velocity, **v** m s⁻¹, of *P* is given by

$$\mathbf{v} = 4t^2\mathbf{i} + (5t - 1)\mathbf{j}$$

Find:

(a) the time after which the particle is moving parallel to the vector **i**

(1)

(b) the acceleration of *P* at time *t* seconds

(2)

(c) the magnitude of **F** when *t* = 1.5.

(3)

(Total for Question 7 is 6 marks)

8   The diagram shows two particles, A of mass 6 kg and B of mass 2.5 kg, connected by a light inextensible string passing over a smooth pulley.

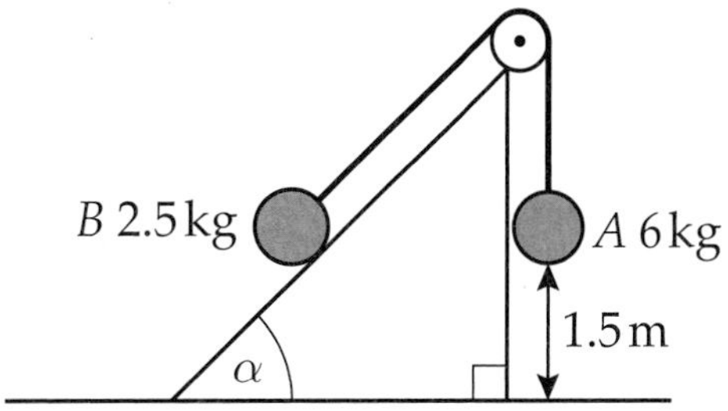

Initially B is held at rest on a rough inclined plane, inclined at an angle α to the horizontal, where $\tan \alpha = \frac{4}{3}$.

The coefficient of friction between B and the inclined plane is $\frac{1}{6}$.

The system is released from rest with A at a height 1.5 m above the ground.

(a)   For the motion until A hits the ground, work out the acceleration of the system.

(7)

(b) A does not rebound when it hits the ground and B continues moving up the inclined plane. Given that B does not reach the pulley, find how much further B travels before coming to instantaneous rest.

(5)

(c)   State how you have used the fact that

   (i)   the string is inextensible

   (ii) the pulley is smooth

   in your calculations.

(2)

Revision Guide
page 175

Problem solving

In parts (a) and (b) you need to think carefully about the forces acting on the particles. Use components of the 2.5 kg mass, both parallel and perpendicular to the inclined plane. The movement is up the plane in both parts, so friction acts down the inclined plane.

Draw clear diagrams and show all the forces acting on the particles.

The question involves resolving forces, using $F = ma$ and the *suvat* formulae.

Hint Q8a

Use $F = ma$ for both particles and solve simultaneously to find a.

Hint Q8b

Once A has hit the ground, the only forces acting on B are its weight components and friction.

Use $F = ma$ and one of the *suvat* formulae.

(Total for Question 8 is 14 marks)

9 A basketball player is taking two free-throw shots. He stands 4.2 m from the centre of the hoop of the basket. The basket is 3 m above the ground and the player releases the basketball from a height of 2 m.

For his first shot, the basketball is projected at an angle of 50° with a speed of 7.5 m s⁻¹.

(a) Show that the shot will be slightly too high to pass through the hoop of the basket.

(6)

For his second shot, the player keeps the angle of projection at 50° but adjusts the speed of the throw. This time, the basketball passes through the centre of the hoop and he scores one point.

(b) Find the new speed of projection.

(6)

....

....

....

....

....

....

....

....

....

....

....

....

....

....

....

(Total for Question 9 is 12 marks)

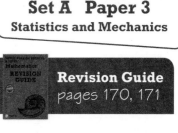

Revision Guide
pages 170, 171

Problem solving

In both parts, you need to establish the time for the basketball to reach the hoop, using the horizontal component of the speed. Then use this information, along with a *suvat* formula, for the vertical motion.

Hint Q9b

You are effectively using the equation of the trajectory. This is a particular case of that concept. The equation for u (the required speed) looks a little complicated, so be careful when isolating the term in u^2, which leads to the solution.

Modelling

You are modelling the basketball as a particle. Always interpret your solutions in the context of the problem.

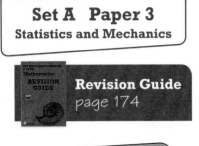

Revision Guide
page 174

Problem solving

Draw a diagram for both parts of this question. Show normal reactions, frictional forces and the weights of the ladder and the man standing on it.

Hint Q10a

Resolve horizontally and vertically and choose a point to take moments about.

Remember, moment = force × perpendicular distance.

Hint Q10b

There is an extra force (kW) acting vertically downwards at A and the man is now standing at the top of the ladder. Use the value of μ from part (a), and resolve and take moments as before. If you consider friction to be limiting, you will be calculating the minimum value of k.

Modelling

The man is modelled as a particle. The ladder is uniform, so its weight will act at its midpoint.

10

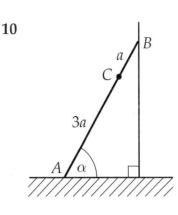

A ladder, AB, of weight W and length 4a, has one end on rough horizontal ground. The coefficient of friction is μ. The other end rests against a smooth vertical wall and the ladder makes an angle α with the ground, where $\tan\alpha = \dfrac{7}{2}$.

A man of weight 6W stands at C on the ladder, where AC = 3a. The ladder is modelled as a uniform rod, lying in a vertical plane, and the man is modelled as a particle. The system is in limiting equilibrium.

(a) Show that $\mu = \dfrac{10}{49}$

(6)

A downward force, equivalent to a weight of kW, is applied at A, enabling the man to climb to the top of the ladder at B. The system is in equilibrium.

(b) Find the range of possible values of k.

(7)

....

....

....

....

....

....

....

....

....

....

....

....

....

(Total for Question 10 is 13 marks)

TOTAL FOR PAPER IS 100 MARKS

Revision Guide
page 87

LEARN IT!

The small angle
approximations are:

$\sin\theta \approx \theta$

$\tan\theta \approx \theta$

$\cos\theta \approx 1 - \dfrac{\theta^2}{2}$

Hint Q1b

Small angle
approximations only
work when the angle is
given in radians.

Paper 1: Pure Mathematics 1

Answer all questions. Write your answers in the spaces provided.

1 (a) Given that θ is small, use the small angle approximations for
$\cos\theta$ and $\sin\theta$ to show that

$$\frac{1 - \cos^2 3\theta}{3\theta \sin 2\theta} \approx \frac{3}{2}$$

(3)

Beth uses $\theta = 5°$ to test the approximation in part (a). This is her
working:

> Using my calculator:
> $$\frac{1 - \cos^2 15°}{15 \times \sin 10°} = 0.0257\ldots$$
> So the approximation is not true for $\theta = 5°$

(b) Identify the mistake in Beth's working, and show that the
approximation is accurate to 1 decimal place.

(2)

(Total for Question 1 is 5 marks)

2 The function f is defined by f: $x \rightarrow |2x - 7|$, $x \in \mathbb{R}$.

(a) Sketch the graph with equation $y = f(x)$, showing the coordinates of the points where the graph cuts or meets the axes.

(2)

(b) Solve $f(x) = 20 + x$

(3)

Revision Guide
page 66

Hint Q2b

There will be two solutions. Consider the positive and negative arguments.

Problem solving

You could use your sketch to check that your solutions look about right.

(Total for Question 2 is 5 marks)

Revision Guide
page 28

Problem solving

This is called the ambiguous case because there are two possible triangles that work with these values. Sketch the triangle then use the sine rule.

LEARN IT!

$$\frac{\sin A}{a} = \frac{\sin B}{b}$$

3 In the triangle ABC, $AB = 8\,\text{cm}$, $AC = 11\,\text{cm}$ and angle $ACB = \frac{\pi}{9}$ radians.

Find the two possible values of angle ABC.

(5)

(Total for Question 3 is 5 marks)

4   $f(x) = e^{x\sqrt{3}} \cos x, \quad \dfrac{-\pi}{2} \leqslant x \leqslant \dfrac{\pi}{2}$

The diagram shows a sketch of the curve C with equation $y = f(x)$

Revision Guide
pages 39/40,
90, 92

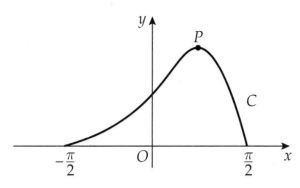

Hint Q4a

Use the product rule, then solve a trig equation for the x-coordinate of P.

(a) Find the x-coordinate of the stationary point P, on C.

    Give your answer as a multiple of π.

(4)

Hint Q4b

You will need to find the values of y and $\dfrac{dy}{dx}$ when $x = 0$.

(b) Find the equation of the tangent to C at the point where $x = 0$.

(3)

$f(x) = e^{x\sqrt{3}} \cos x, \quad \dfrac{-\pi}{2} \leqslant x \leqslant \dfrac{\pi}{2}$

(Total for Question 4 is 7 marks)

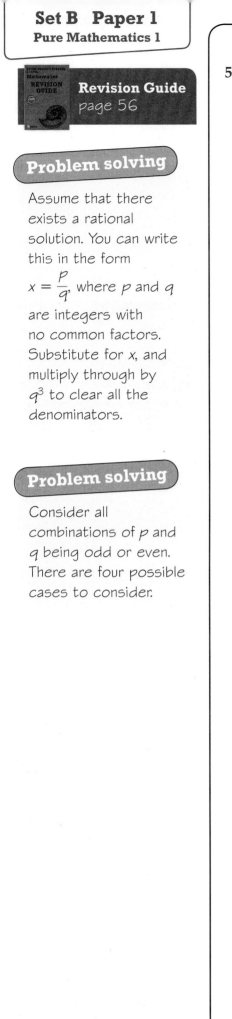

Revision Guide
page 56

Problem solving

Assume that there exists a rational solution. You can write this in the form $x = \dfrac{p}{q}$, where p and q are integers with no common factors. Substitute for x, and multiply through by q^3 to clear all the denominators.

Problem solving

Consider all combinations of p and q being odd or even. There are four possible cases to consider.

5   Use proof by contradiction to show that there are no rational numbers which satisfy the equation

$$x^3 - 3x + 1 = 0$$

(5)

(Total for Question 5 is 5 marks)

6 A circle intersects the *x*-axis at the points *A*(2, 0) and *B*(8, 0).
The *y*-axis is a tangent to the circle at *C*.

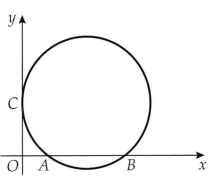

(a) Work out the equation of the circle.

(4)

(b) The point *E* is on the circle vertically above *B*. Find the equation
of the tangent at the point *E* in the form $ax + by = c$.

(5)

Revision Guide
pages 20, 21

Hint Q6a

The midpoint of *AB* is
the *x*-coordinate of the
centre.

Use Pythagoras to find
the *y*-coordinate of the
centre of the circle.

Hint Q6b

Find the coordinates of
E. Find the gradient of
the line that connects
E and the centre of
the circle. Use this to
find the gradient of the
tangent and then use
$y - y_1 = m(x - x_1)$ with
the coordinates of *E* to
find the equation of the
tangent.

LEARN IT!

The gradient of the
tangent is the negative
reciprocal of the radius.

(Total for Question 6 is 9 marks)

7 Solve, for $-\pi < x < \pi$, the equation

$$6\sin^2 x + \cos x - 4 = 0$$

giving your answers to 2 decimal places.

(5)

Revision Guide
page 32

Hint

$\sin^2 x = 1 - \cos^2 x$.
Simplify the resulting
quadratic and factorise.

Watch out!

There are four solutions
in total.

(Total for Question 7 is 5 marks)

Revision Guide
page 75

Hint Q8a

Write the expression as $(x + 1)(2 - 3x)^{-2}$ then take a common factor of 2 out of the second bracket.

When 2 is taken out as a common factor it becomes 2^{-2}. Expand the second bracket to the x^2 term, then multiply by the first bracket.

LEARN IT!

Remember:
$$(a + bx)^n = a^n\left(1 + \frac{bx}{a}\right)^n$$
and the expression is valid for
$$\left|\frac{bx}{a}\right| < 1 \text{ or}$$
$$|x| < \frac{a}{b}$$

8 (a) Expand

$$\frac{x + 1}{(2 - 3x)^2}$$

in ascending powers of x, up to and including the term in x^2, giving each term as a simplified fraction.

(5)

(b) State the range of values of x for which the expansion converges.

(1)

....

....

....

....

....

....

....

....

....

....

....

....

....

....

....

....

....

....

(Total for Question 8 is 6 marks)

9   A liquid is being heated. At time t seconds, the temperature of the liquid is $\theta\,°C$.

The rate of increase of the temperature of the liquid is modelled by the differential equation

$$\frac{d\theta}{dt} = k(170 - 2\theta)$$

where k is a positive constant.

(a) Given that $\theta = 20$ when $t = 0$, show that $\theta = A - B\,e^{-2kt}$ where A and B are integers to be found.

(6)

(b) Given that $k = 0.004$, find the time for the temperature of the liquid to reach $70\,°C$.

(2)

Revision Guide
pages 41, 97, 109

Hint Q9a

Separate the variables then integrate. Don't forget the constant of integration, which must be evaluated.

LEARN IT!

$$\int \frac{1}{a + bx}dx$$
$$= \frac{1}{b}\ln(a + bx) + c$$

(Total for Question 9 is 8 marks)

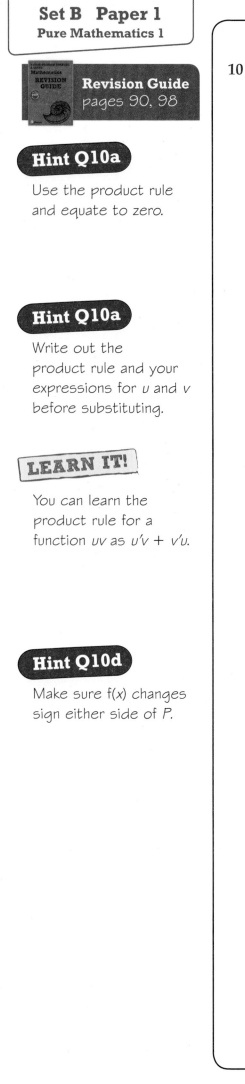

Revision Guide
pages 90, 98

Hint Q10a

Use the product rule and equate to zero.

Hint Q10a

Write out the product rule and your expressions for u and v before substituting.

LEARN IT!

You can learn the product rule for a function uv as $u'v + v'u$.

Hint Q10d

Make sure f(x) changes sign either side of P.

10

The diagram shows a sketch of the graph of

$$f(x) = 9x^2 e^{2x} - 4, \ x \in \mathbb{R}$$

(a) Find the exact coordinates of the turning points on this curve.

(5)

(b) Show that the equation $f(x) = 0$ can be written in the form

$$x = \pm\frac{2}{3}e^{-x}$$

(1)

The equation f(x) has a root α, where $\alpha = 0.4$, to 1 decimal place.

(c) Starting with $x_0 = 0.4$, use the iterative formula

$$x_{n+1} = \frac{2}{3}e^{-x_n}$$

to calculate the values of x_1, x_2 and x_3, giving your answers to 4 decimal places.

(3)

(d) Give an accurate estimate to 2 decimal places of the x-coordinate of P, justifying your answer.

(2)

....

....

....

....

....

....

....

....

....

....

....

....

....

....

....

....

....

....

....

....

....

....

....

(Total for Question 10 is 11 marks)

Revision Guide
page 8

Hint Q11a

The information given
in the question tells
you the coordinates of
the turning point and
one other point on the
parabola.

Modelling

You can give any
reasonable answer but
make sure you refer
to the context of the
question.

11 The diagram shows the trajectory of an arrow that is fired
horizontally from the top of a castle.

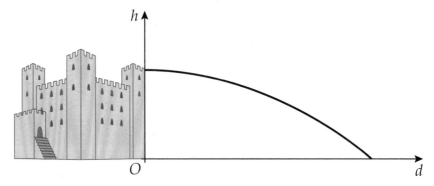

The point of projection is 20 m above ground level, and the arrow hits
the ground a horizontal distance of 100 m from the base of the castle.

(a) Given that the trajectory of the arrow can be modelled as a
parabola, find an equation for the height of the arrow above the
ground, h, in terms of the horizontal distance travelled, d.

Give your answer in the form $h = ad^2 + bd + c$, and specify a
suitable range of values for d.

(4)

(b) State one limitation of the model.

(1)

(Total for Question 11 is 5 marks)

12 $\displaystyle\sum_{r=1}^{k}(3 + \frac{1}{2}r) = 105$

Find the value of k.

(5)

Revision Guide
pages 68, 72

Problem solving

Write out the first few terms of the arithmetic series with general term $3 + \frac{1}{2}r$, then write an expression for the sum to k terms.

Hint

k must be a positive integer.

....

....

....

....

....

....

....

....

....

....

....

....

....

....

....

....

....

....

....

(Total for Question 12 is 5 marks)

Revision Guide
pages 90, 103, 108

Hint Q13a

Two of the answers are given so you can check that you are doing the correct operation on your calculator. Check that your values look to fit the shape of the curve in the diagram.

Hint Q13b

The trapezium rule is
$\frac{1}{2}h[(y_0 + y_n) + 2(y_1 + y_2 + ... + y_{n-1})]$
where $h = \frac{b - a}{n}$.

This is given in the formulae book that you are provided with in your examination.

Hint Q13c

Differentiate $\cot x$ to get dx in terms of x and du. Substitute $\cot x$ and dx for terms with u and cancel out the cosec^2 terms.

Find u when $x = \frac{\pi}{6}$ and $x = \frac{\pi}{3}$, and substitute in for the limits. Then integrate with respect to u and find the exact value by manipulating the surds.

13

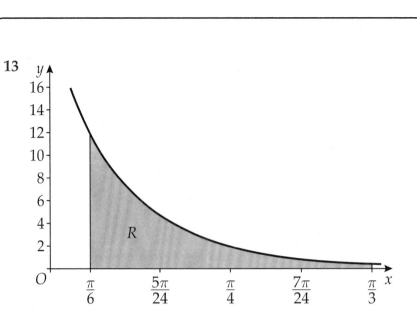

The diagram shows the graph of the curve with equation

$$y = \operatorname{cosec}^2 x \cot^2 x, \quad \frac{\pi}{6} \leqslant x \leqslant \frac{\pi}{3}$$

The finite region R is bounded by the lines $x = \frac{\pi}{6}$, $x = \frac{\pi}{3}$, the x-axis and the curve as shown.

(a) Complete the table with the values of y corresponding to $x = \frac{5\pi}{24}, \frac{\pi}{4}$ and $\frac{7\pi}{24}$

x	$\frac{\pi}{6}$	$\frac{5\pi}{24}$	$\frac{\pi}{4}$	$\frac{7\pi}{24}$	$\frac{\pi}{3}$
y	12				0.44444

(2)

(b) Use the trapezium rule with the values in the table to find an approximate value for the area of R, giving your answer to 3 significant figures.

(4)

(c) By using the substitution $u = \cot x$ and integrating, find the exact value of the area of R.

(7)

(Total for Question 13 is 13 marks)

Revision Guide
pages 84, 85

Hint Q14b, c

Use the result from part (a) and consider the significance of $\cos(\theta + \alpha)$ when the cabin is at its highest.

Hint Q14d

Find the two values of t and work out the length of time between them.

14 (a) Express $35\cos\theta - 12\sin\theta$ in the form $R\cos(\theta + \alpha)$, where $R > 0$ and $0 < \alpha < \dfrac{\pi}{2}$.
State the value of R and give the value of α to 4 decimal places.

(3)

An amusement park has a big-wheel ride at the entrance.
The height above the ground of one of the cabins is modelled by the equation

$$h = 39 - 35\cos 0.25t + 12\sin 0.25t, \quad t > 0$$

where h is the height of the cabin, in metres, and t is the number of minutes after a visitor boards the cabin. The angles are measured in radians.

Find:

(b) the maximum height reached by the cabin

(2)

(c) the time taken for a cabin to reach its maximum height.

(2)

The best panoramic views are obtained when the cabins are 50 m or more above the ground.

(d) Calculate the number of minutes that a cabin is at least 50 m above the ground in each revolution of the big wheel.

(4)

....

....

....

....

....

....

....

....

....

....

....

....

....

....

....

....

....

....

....

....

(Total for Question 14 is 11 marks)

TOTAL FOR PAPER IS 100 MARKS

Revision Guide
page 60

Hint Q1a

Make sure you get the order of the functions correct, and don't forget to simplify your answers.

Hint Q1b

A value of x can only be the solution to $fg(x) = gf(x)$ if it lies within the domain of both functions.

Paper 2: Pure Mathematics 2

Answer all questions. Write your answers in the spaces provided.

1 $f(x) = e^{2x} + 2, \quad x \in \mathbb{R}$

$g(x) = \ln(x - 2), \quad x > 2, x \in \mathbb{R}$

(a) Write down the composite functions

(i) $fg(x)$

(ii) $gf(x)$

simplifying your answers.

(3)

(b) Hence find the solution to $fg(x) = gf(x)$, and justify that it is unique.

(2)

....

....

....

....

....

....

....

....

....

....

....

....

....

....

....

(Total for Question 1 is 5 marks)

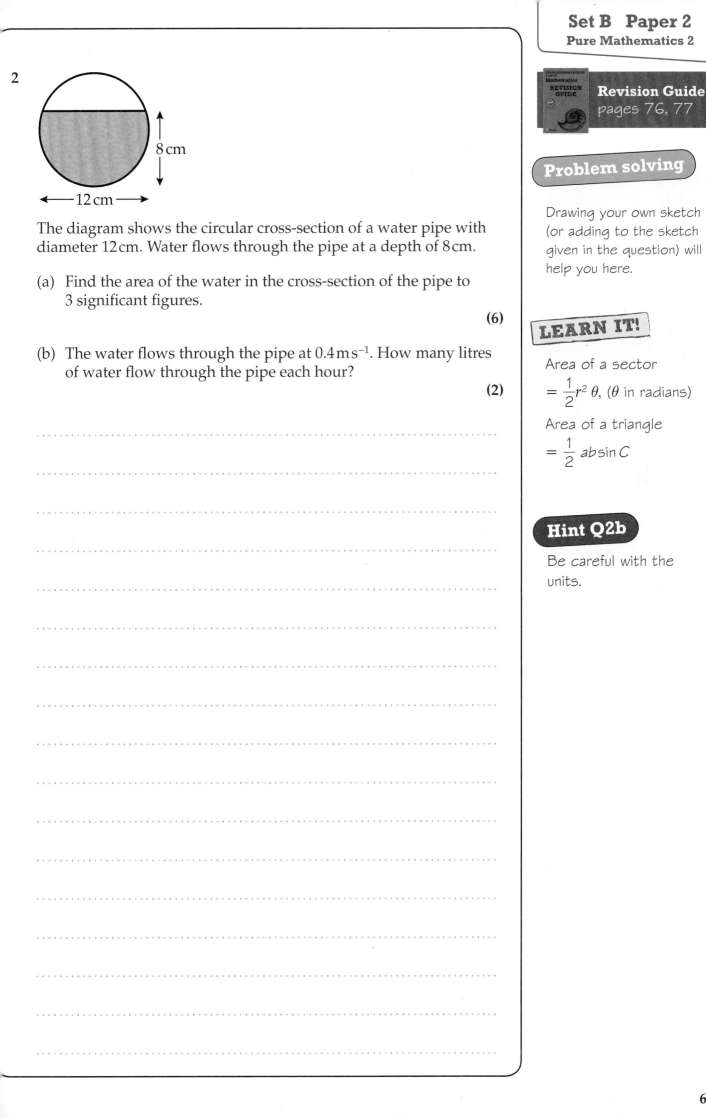

2

8 cm

12 cm

The diagram shows the circular cross-section of a water pipe with diameter 12 cm. Water flows through the pipe at a depth of 8 cm.

(a) Find the area of the water in the cross-section of the pipe to 3 significant figures.

(6)

(b) The water flows through the pipe at $0.4\,\mathrm{m\,s^{-1}}$. How many litres of water flow through the pipe each hour?

(2)

Revision Guide
pages 76, 77

Problem solving

Drawing your own sketch (or adding to the sketch given in the question) will help you here.

LEARN IT!

Area of a sector
$= \dfrac{1}{2}r^2\,\theta$, ($\theta$ in radians)

Area of a triangle
$= \dfrac{1}{2}\,ab\sin C$

Hint Q2b

Be careful with the units.

(Total for Question 2 is 8 marks)

3 (a) A geometric series has first term a and common ratio r.

Prove that the sum, S, of the first n terms of the series is

$$\frac{a(1 - r^n)}{1 - r}$$

(4)

An author will be paid a salary of £27000 for the year 2018.
Her contract promises a 4% increase in salary every year, the first
increase being given in 2019, so that these annual salaries form a
geometric sequence.

(b) Find, to the nearest £100, the salary in the year 2022.

(2)

The author intends to retire in 2040.

(c) Find, to the nearest £1000, the total amount of salary the author
will receive in the period from 2018 until she retires at the end
of 2040.

(3)

(Total for Question 3 is 9 marks)

Revision Guide
pages 69, 70

LEARN IT!

You need to learn the proof for the sum of a geometric series (and the proof for the sum of an arithmetic series).

Hint Q3b

The nth term of a geometric series is ar^{n-1} where a is the first term, r is the common ratio and n is the term number. The term number for 2022 is $n = 5$.

Problem solving

Series questions will often have a practical application. Geometric series are usually used with financial problems or population growth problems.

Revision Guide
page 87

LEARN IT!

The formula for $\cos(A \pm B)$ is given in the formulae book as is the first line of the proof.

You need to be able to differentiate $\sin x$ and $\cos x$ from first principles.

4   Given that x is measured in radians, prove, from first principles, that the derivative of $\cos x$ is $-\sin x$.

You may assume the formula for $\cos (A \pm B)$ and that as $h \to 0$,

$$\frac{\sin h}{h} \to 1 \text{ and } \frac{\cos h - 1}{h} \to 0$$

(5)

(Total for Question 4 is 5 marks)

5 $f(x) = 2x^3 - 3x^{\frac{3}{2}} - 4, \quad x > 0$

(a) The root α of the equation $f(x) = 0$ lies in the interval $[1.7, 1.8]$

Taking 1.8 as a first approximation to α, apply the Newton–Raphson process twice to $f(x)$ to show that $\alpha \approx 1.768$ to 3 decimal places.

(6)

(b) Find the x-coordinate of the stationary point on $f(x)$.

(3)

Revision Guide
page 99

Hint Q5a

The Newton–Raphson formula is given in the formulae book. You can program the formula into your calculator once you are in the examination but not beforehand. If you are not using a program then don't round your answers until the very end. The first step is to differentiate f(x).

$f(x) = 2x^3 - 3x^{\frac{3}{2}} - 4, \quad x > 0$

(Total for Question 5 is 9 marks)

Revision Guide
pages 82, 83

Hint

You will need to use the addition formulae and the double angle formulae as well as $\tan\theta = \dfrac{\sin\theta}{\cos\theta}$

Hint

Write $\cos 3\theta$ as $\cos(2\theta + \theta)$ then use $\cos 2\theta = 1 - 2\sin^2\theta$ later in the proof.

Problem solving

To prove an identity, start with one side, then manipulate the expression using known identities until it matches the other side.

6   Prove that

$$\frac{\sin2\theta\,\tan\theta}{\cos3\theta - \cos\theta} \approx \frac{-1}{2\cos\theta}, \; \theta \neq \frac{n\pi}{2}, \quad n \in \mathbb{Z}$$

(5)

....

....

....

....

....

....

....

....

....

....

....

....

....

....

....

....

....

....

....

....

(Total for Question 6 is 5 marks)

7 The rate of change of y with respect to x is proportional to $\ln x$.
When $y = 0$, $x = 0$ and when $y = 6$, $x = 2$.

Find y when $x = 8$. Write your answer in the form

$$\frac{A\ln 2 - B}{\ln 2 - C}$$

where A, B and C are integers to be found.

(8)

Revision Guide
pages 97, 104

LEARN IT!

$\int k \ln x\, dx$ is a standard integral proof.

Always set $u = \ln x$ when using integration by parts.

Hint

Once the integration is done, substitute in the given values to find c and k. Then substitute $x = 8$ and manipulate into the required form.

(Total for Question 7 is 8 marks)

Revision Guide
pages 48, 49, 50

Hint Q8a

Take natural logs of both sides. Don't round your answers until the very end.

Hint Q8b

Use the rules of logarithms, leading to a quadratic equation in y.

8 (a) Solve $3^{2x-1} = 5^{3x+1}$ giving your answer to 3 significant figures.

(3)

(b) Find the values of y such that

$$\log_3(9y-5) + 2\log_3 2 - \log_3 y - \log_3(y+1) = 2, \quad y > \frac{5}{9}, y \in \mathbb{R}$$

(6)

(Total for Question 8 is 9 marks)

9 $f(x) = \dfrac{16}{(1-x)^2(3+x)}$

(a) Express $f(x)$ in partial fractions.

(4)

(b) Hence, or otherwise, find the series expansion of $f(x)$, in ascending powers of x, up to and including the term in x^2. Simplify each term.

(8)

Revision Guide
pages 24, 58

Hint Q9a

There is a repeated factor in the denominator so you must have denominators of $(1-x)$ and $(1-x)^2$.

$f(x) = \dfrac{A}{1-x} + \dfrac{B}{(1-x)^2} + \dfrac{C}{3+x}$

Either use sensible substitutions or equate coefficients to find A, B and C.

Hint Q9b

Always expand one power further than you are asked for in case dividing or multiplying includes terms you weren't expecting. You can always discard them at the end if you don't need them.

Hint Q9b

$(3+x)^{-1}$ is the awkward expansion here as you need to take a common factor of 3 out first to get $3^{-1}(1 + \frac{1}{3}x)^{-1}$.

(Total for Question 9 is 12 marks)

10 A curve has equation

$$y^2 + 4y = e^x \sin^2 x \quad \text{for } 0 < x < 2\pi$$

Find the values of x where $\dfrac{\mathrm{d}y}{\mathrm{d}x} = 0$.

(7)

....

....

....

....

....

....

....

....

....

....

....

....

....

....

....

....

....

....

....

....

(Total for Question 10 is 7 marks)

Revision Guide
page 94

Hint

There are three different types of differentiation here. The left-hand side needs to be differentiated implicitly, that is, differentiate with respect to y and then put $\dfrac{\mathrm{d}y}{\mathrm{d}x}$ next to each term.

The right-hand side is a product rule and then $\sin^2 x$ can be differentiated using a substitution (although it's likely that you will be able to do this in your head).

LEARN IT!

The derirative of e^x is e^x.

Hint

Once the differentiation is complete, substitute in $\dfrac{\mathrm{d}y}{\mathrm{d}x} = 0$ and solve for x.

LEARN IT!

$e^x \neq 0$ because the graph of $y = e^x$ never crosses the x-axis.

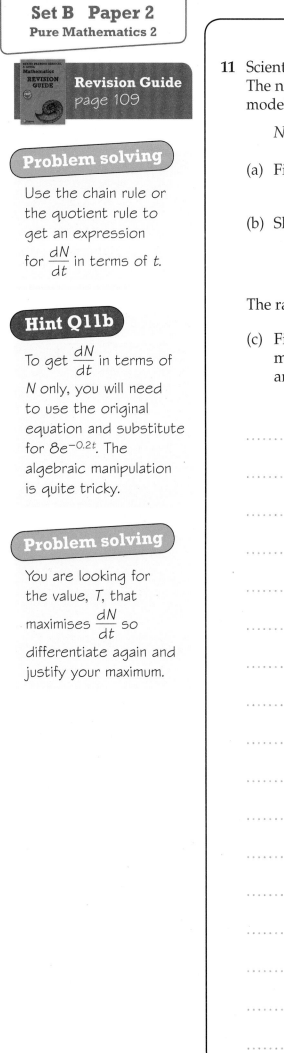

Revision Guide
page 109

Problem solving

Use the chain rule or the quotient rule to get an expression for $\frac{dN}{dt}$ in terms of t.

Hint Q11b

To get $\frac{dN}{dt}$ in terms of N only, you will need to use the original equation and substitute for $8e^{-0.2t}$. The algebraic manipulation is quite tricky.

Problem solving

You are looking for the value, T, that maximises $\frac{dN}{dt}$ so differentiate again and justify your maximum.

11 Scientists are studying the number of bank voles in a woodland. The number, N, at time t months after the start of the study is modelled by the equation

$$N = \frac{660}{3 + 8\,e^{-0.2t}}, \, t \geqslant 0, t \in \mathbb{R}$$

(a) Find the number of bank voles at the start of the study.

(1)

(b) Show that the rate of increase, $\frac{dN}{dt}$, is given by

$$\frac{dN}{dt} = \frac{220N - N^2}{1100}$$

(5)

The rate of increase in bank voles is a maximum after T months.

(c) Find the value of T as predicted by this model and state the maximum number of bank voles in the woodland. Justify your answers.

(6)

(Total for Question 11 is 12 marks)

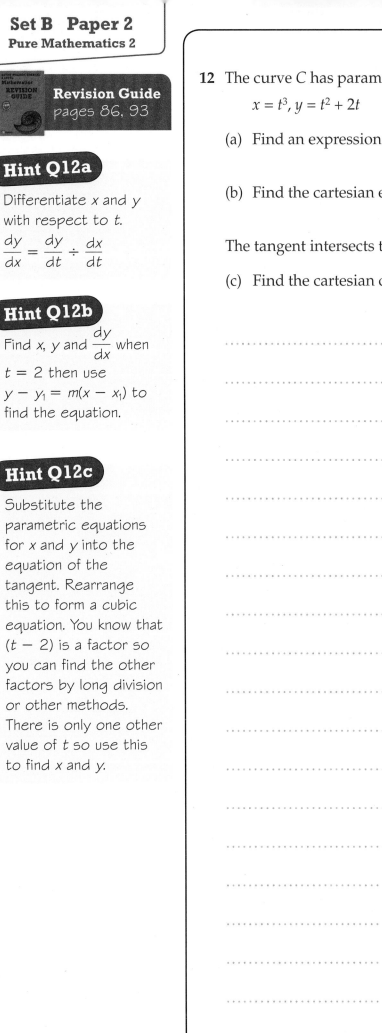

Revision Guide
pages 86, 93

Hint Q12a

Differentiate x and y with respect to t.

$$\frac{dy}{dx} = \frac{dy}{dt} \div \frac{dx}{dt}$$

Hint Q12b

Find x, y and $\frac{dy}{dx}$ when $t = 2$ then use $y - y_1 = m(x - x_1)$ to find the equation.

Hint Q12c

Substitute the parametric equations for x and y into the equation of the tangent. Rearrange this to form a cubic equation. You know that $(t - 2)$ is a factor so you can find the other factors by long division or other methods. There is only one other value of t so use this to find x and y.

12 The curve C has parametric equations

$$x = t^3, y = t^2 + 2t$$

(a) Find an expression for $\frac{dy}{dx}$ in terms of t.

(2)

(b) Find the cartesian equation of the tangent to C when $t = 2$.

(3)

The tangent intersects the curve C again at the point D.

(c) Find the cartesian coordinates of D.

(6)

(Total for Question 12 is 11 marks)

TOTAL FOR PAPER IS 100 MARKS

Revision Guide
pages 136, 137

Hint Q1a

You can find PMCCs from given data values using your calculator.

Hint Q1b

This is a one-tailed test. Remember to state your conclusion in the context of the question.

Watch out!

You must state your hypotheses in terms of the population PMCC, *p*, and show a comparison between the *r*-value from the table and the *r*-value you have calculated.

Paper 3: Statistics and Mechanics
SECTION A: STATISTICS

Answer all questions. Write your answers in the spaces provided.

1   A random sample of 10 days in Perth, Western Australia, in September 2015, was chosen, and for each day a mean daily temperature and a mean daily pressure reading was taken.

Mean daily temperature (°C)	11.8	12.3	22.5	11.2	12.1	13.6	17.7	14.2	20.1	15.7
Mean daily pressure (hPa)	1025	1028	1015	1018	1025	1026	1022	1029	1018	1022

   (a) Calculate the product moment correlation coefficient for these data.

   (1)

   (b) Stating your hypotheses clearly, test, at the 2.5% level of significance, whether or not the product moment correlation coefficient for these data is less than zero.

   (3)

   (c) Give an interpretation of the value 2.5% in your hypothesis test.
   (1)

   (d) Suggest how you could make better use of the large data set to investigate the relationship between the mean daily temperature and the mean daily pressure.

   (1)

   (Total for Question 1 is 6 marks)

2   A box, *C*, contains 7 counters of which 4 are red and 3 are white.

A box, *D*, contains 6 counters of which 2 are red and 4 are white.

A counter is drawn at random from *C* and placed into *D*. A second counter is drawn at random from *C* and placed into *D*. A third counter is then drawn at random from the counters in *D*.

(a) Draw a tree diagram to show this situation, showing all probabilities.

(4)

(b) Find the probability that a white counter is drawn from *D*.

(2)

(c) Given that 2 of the 3 counters drawn from *D* are the same colour, find the probability that they are '2 red and 1 white'.

(5)

Revision Guide
page 129

Hint Q2c

When you are asked for '2 red and 1 white', this can be in any order, so make sure you consider all possible combinations.

Hint Q2c

$$P(B \mid A) = \frac{P(A \cap B)}{P(A)}$$

(Total for Question 2 is 11 marks)

Revision Guide
pages 140, 142, 143

Hint Q3c

Use the inverse normal distribution function on your calculator.

Hint Q3d

Draw a sketch of the distribution. You will see that since $P(Y < 33) = 0.2$, the z-value corresponding to it will be negative. Use the percentage points of the normal distribution table and set up two equations in μ and σ and solve simultaneously.

3   In an aptitude test, the scores, X, are normally distributed with mean 48.5 and standard deviation 9.7

(a) Find $P(X < 52)$

(1)

(b) Find $P(45 < X < 55)$

(1)

(c) A score of k or better was attained by the top 10% of the candidates sitting this aptitude test.

Find the value of k.

(2)

In a second aptitude test, the scores, Y, are normally distributed with mean μ and standard deviation σ.

(d) Given that $P(Y < 33) = 0.2$ and $P(Y < 54) = 0.85$, find the values of μ and σ.

(7)

(Total for Question 3 is 11 marks)

4 Given that $P(A) = 0.45$, $P(B) = 0.3$ and $P(B \mid A) = 0.2$, find:

(a) $P(A \cap B)$

(2)

(b) $P(A' \cap B)$

(1)

Event C has $P(C) = 0.35$ and $P(B \mid C) = 0.4$

Events A and C are mutually exclusive.

(c) Find $P(B \cap C)$.

(2)

(d) Draw a Venn diagram to illustrate the events A, B and C, giving the probabilities for each region.

(5)

Set B Paper 3
Statistics and Mechanics

Revision Guide
pages 127, 128, 138

Hint Q4a

Use the conditional probability definition.

Hint Q4c

Use the conditional probability definition.

Hint Q4d

Don't forget to give the probability for the region outside A, B and C.

....

....

....

....

....

....

....

....

....

....

....

....

....

....

....

....

(Total for Question 4 is 10 marks)

Revision Guide
page 145

Problem solving

There are many stages in answering this question. You will need to use the percentage points of the normal distribution table to find the mean thickness for the first machine. Only then can you find the probability in part (a).

Then, think about an appropriate probability distribution to use in part (b).

Hint Q5c

Work out the variance for the sample $\left(\dfrac{\sigma^2}{n}\right)$ before you use a hypothesis test.

Modelling

Always interpret your conclusions in the context of the question.

5 A machine produces metal sheets of thickness T mm, where T is normally distributed with standard deviation 0.15 mm.

Sheets less than 5.5 mm thick or more than 6.0 mm thick **cannot** be used.

Given that 2.5% of sheets have a thickness of less than 5.45 mm:

(a) find the probability than a randomly chosen metal sheet **can** be used.

(5)

Fifteen metal sheets are chosen at random.

(b) Find the probability that fewer than 4 of them **cannot** be used.

(2)

Another machine also produces metal sheets of thickness X mm, where X is normally distributed with standard deviation 0.18 mm.

A random sample of 25 sheets produced by this machine is taken and the sample mean thickness was found to be 5.88 mm.

(c) Stating your hypotheses clearly, and using a 2% level of significance, test whether the mean thickness of metal sheets produced by this machine is greater than 5.8 mm.

(5)

(Total for Question 5 is 12 marks)

Revision Guide
page 178

Hint

You need to split the time into two parts, 0 to 5π and 5π to 30, and integrate the appropriate expression. Add the answers for the final displacement.

LEARN IT!

Integrate the velocity to find the displacement. If you want to find the displacement after a fixed period of time you can use definite integration.

SECTION B: MECHANICS

Answer all questions. Write your answers in the spaces provided.

Unless otherwise indicated, whenever a numerical value of g is required, take $g = 9.8\,\text{m}\,\text{s}^{-2}$ and give your answer to either 2 significant figures or 3 significant figures.

6   A particle, P, is moving in a straight line. At time t seconds, $t \geqslant 0$, the velocity of P, $v\,\text{m}\,\text{s}^{-1}$, is given by

$$v = \begin{cases} 2 + \sin 0.2t, & 0 < t \leqslant 5\pi \\ 2\cos 0.4t, & 5\pi \leqslant t \leqslant 10\pi \end{cases}$$

Find the displacement of P from its starting position after 30 seconds.

(8)

(Total for Question 6 is 8 marks)

7

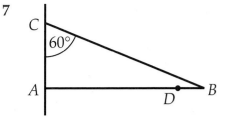

A uniform rod AB, of mass $4m$ and length $6a$, is held in a horizontal position with the end A against a rough vertical wall. One end of a light inextensible string is attached to the rod at B and the other end to a point C on the wall vertically above A. A particle of mass $3m$ is attached to the rod at D, where $AD = 5a$. The angle between the wall and the string at C is $60°$. The rod is in equilibrium in a vertical plane perpendicular to the wall.

(a) Find the tension in the string.

(4)

The coefficient of friction between the rod and the wall is μ. Given that the rod is about to slip,

(b) find the value of μ.

(6)

Problem solving

Draw a diagram and mark on it all the forces acting on the rod.

Hint Q7a

Take moments about an appropriate point to find the tension in the string.

Hint Q7b

The rod is about to slip, so friction is limiting. Do not round any answers prematurely. Keep at least 4 decimal places or work with surds.

(Total for Question 7 is 10 marks)

8

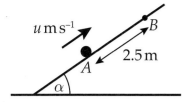

Revision Guide
page 169

A particle has a mass of 3 kg and is projected from a point A, with a speed of $u\,\mathrm{m\,s^{-1}}$, up a line of greatest slope of a rough inclined plane. The plane makes an angle α with the horizontal where $\tan\alpha = \dfrac{4}{3}$.

The coefficient of friction between the particle and the plane is $\dfrac{1}{6}$.

Given that the particle comes to instantaneous rest at B, where $AB = 2.5\,\mathrm{m}$,

(a) find the value of u.

(7)

The particle then moves back down the inclined plane.

(b) Find the speed of the particle as it passes through A.

(4)

Problem solving

Draw a diagram and mark on it all the forces acting on the particle.

Hint Q8a

Use a combination of $F = ma$ and a suvat formula to work out u.

Hint Q8b

Now that the particle is moving down the slope, friction will act upwards to oppose the motion.

(Total for Question 8 is 11 marks)

9 In this question, **i** and **j** are horizontal unit vectors due east and due north respectively.

At time t seconds, where $t \geqslant 0$, a particle, P, is moving in a horizontal plane with acceleration

$a = [(3t + 2)\mathbf{i} + (t + 4)\mathbf{j}]\,\mathrm{m\,s^{-2}}$

When $t = 2$, the velocity of P is $(6\mathbf{i} + 9\mathbf{j})\,\mathrm{m\,s^{-1}}$.

Find:

(a) the velocity of P at time t seconds

(5)

(b) the speed of P when it is moving on a bearing of 045°.

(5)

(Total for Question 9 is 10 marks)

Revision Guide
pages 176, 177, 179

Hint Q9a

Use calculus and don't forget to include a constant of integration vector.

Hint Q9b

What will be the connection between the **i** and **j** components of the velocity when P is moving on a bearing of 045°?

You are asked for the speed of P so don't leave your answer in vector form.

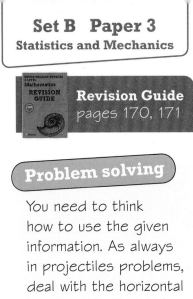

Revision Guide
pages 170, 171

Problem solving

You need to think how to use the given information. As always in projectiles problems, deal with the horizontal motion and vertical motion separately.

Hint Q10a

Use the 15 m in a *suvat* formula to find *u*.

Hint Q10b

Consider the vertical motion from *P* to *R*, then the horizontal motion. What must you calculate to connect both of these?

Modelling

If you have to criticise a model, consider the real-world situation. Make sure you give two different reasons.

10

$u\,\mathrm{m\,s^{-1}}$

Q

15 m

P 54°

42 m

O R

A ball is projected from a point P which is 42 m above O.

The speed of projection is $u\,\mathrm{m\,s^{-1}}$ and the angle of projection is 54° to the horizontal.

The highest point of its path, Q, is 15 m above P, and the ball lands on the ground at R.

The ball is modelled as a particle moving freely under gravity in a vertical plane, and the ground is modelled as level.

Taking $g = 10\,\mathrm{m\,s^{-2}}$, find:

(a) the value of u

(3)

(b) the distance OR.

(6)

(c) State two limitations of this model.

(2)

....

....

....

....

....

....

....

....

....

....

....

....

....

....

....

....

....

....

(Total for Question 10 is 11 marks)

TOTAL FOR PAPER IS 100 MARKS

Paper 1: Pure Mathematics 1

Revision Guide
page 39

Answer **all** questions. Write your answers in the spaces provided.

1   Find the set of values for which

$$f(x) = x^3 - 5x^2 + 3x + 4$$

is a decreasing function.

(4)

Hint

Start by differentiating.

LEARN IT!

$f(x)$ is decreasing on an interval $[a, b]$ if $f'(x) \leqslant 0$ for all values of x in that interval.

Hint

You can give your answers as an inequality or using set notation.

$f(x) = x^3 - 5x^2 + 3x + 4$

$f'(x) = 3x^2 - 10x + 3$ ✓

$ = (3x - 1)(x - 3)$ ✓

Stationary points at $x = \dfrac{1}{3}$ and $x = 3$ on a cubic curve.

Test $x = 1$: $f'(x) = -4$ ✓

So for $\dfrac{1}{3} \leqslant x \leqslant 3$, $f(x)$ is a decreasing function. ✓

(Total for Question 1 is 4 marks)

1

Revision Guide
pages 58, 60
62

Hint Q2a

Evaluate f(4) then use the answer as your input for f.

Hint Q2b

Use partial fraction techniques and compare coefficients.

Hint Q2c

Consider the values of f(x) when $x = 2$ and when $x \to \infty$. You can use your answer to part (b) to work out the value of f(x) as $x \to \infty$.

Hint Q2d

The domain of the inverse function is the same as the range of the original function.

2 $f(x) = \dfrac{5x + 2}{2x - 1}$, $x \geqslant 2$

(a) Find ff(4)

(2)

(b) Show that f(x) can be written in the form $A + \dfrac{B}{2x - 1}$, where A and B are constants to be found.

(2)

(c) Hence, or otherwise, state the range of f.

(1)

(d) Find $f^{-1}(x)$, stating its domain.

(3)

(a) $f(4) = \dfrac{5(4) + 2}{2(4) - 1} = \dfrac{22}{7}$

$ff(4) = \dfrac{5\left(\frac{22}{7}\right) + 2}{2\left(\frac{22}{7}\right) - 1}$ ✓

$= \dfrac{124}{37}$ ✓

(b) $\dfrac{5x + 2}{2x - 1} = A + \dfrac{B}{2x - 1}$

$5x + 2 = A(2x - 1) + B$

$5 = 2A$, so $A = 2.5$

$2 = -A + B$, so $B = 4.5$

$\dfrac{5x + 2}{2x - 1} = 2.5 + \dfrac{4.5}{2x - 1}$ ✓✓

(c) Range of f is $2.5 < f(x) \leqslant 4$ ✓

(d) Let $y = f^{-1}(x)$, then $f(y) = x$

$\dfrac{5y + 2}{2y - 1} = x$ ✓

$5y + 2 = 2xy - x$

$x + 2 = 2xy - 5y$

$= y(2x - 5)$

So $y = \dfrac{x + 2}{2x - 5}$, i.e. $f^{-1}(x) = \dfrac{x + 2}{2x - 5}$ ✓

Domain is $2.5 < y \leqslant 4$ ✓

(Total for Question 2 is 8 marks)

2

3   The diagram shows the graph of

$$y = 24x - 8x^{\frac{3}{2}}$$

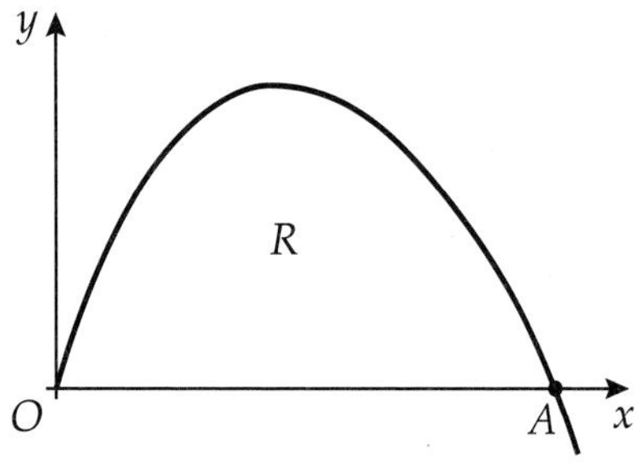

Revision Guide page 45

Hint Q3a

Solve $24x - 8x^{\frac{3}{2}} = 0$

Hint Q3b

Area $= \displaystyle\int_{0}^{a} (24x - 8x^{\frac{3}{2}})\, dx$

The graph crosses the x-axis at A.

(a) Find the value of the x-coordinate at A.

(2)

(b) Find the area, R, bounded by the curve and the x-axis between the origin, O, and point A.

(4)

(a) $y = 24x - 8x^{\frac{3}{2}}$

$24x - 8x^{\frac{3}{2}} = 0$

$24x = 8x^{\frac{3}{2}}$ ✔

$x^{\frac{1}{2}} = 3$, giving $x = 9$ ✔

(b) Area of $R = \displaystyle\int_{0}^{9} (24x - 8x^{\frac{3}{2}})\, dx$ ✔

$= \left[12x^2 - \dfrac{16}{5}x^{\frac{5}{2}} \right]_{0}^{9}$ ✔

$= \left(972 - \dfrac{16}{5} \times 243 \right) - (0)$ ✔

$= 194.4$ ✔

(Total for Question 3 is 6 marks)

Revision Guide
pages 39, 40, 96

Hint Q4b

The stationary points occur when $\frac{dy}{dx} = 0$

Hint Q4c

At a point of inflexion, $\frac{d^2y}{dx^2} = 0$

Problem solving

$\frac{d^2y}{dx^2} = 0$ doesn't guarantee a point of inflexion. You also need the sign of $\frac{d^2y}{d^2x}$ to change on either side of that point.

4 A curve, C, has equation

$$y = x^3 + 6x^2 + 9x + 5$$

(a) Find $\frac{dy}{dx}$ and $\frac{d^2y}{dx^2}$

(3)

(b) Verify that C has a stationary point when $x = -3$.

(2)

(c) Determine the nature of this stationary point, giving a reason for your answer.

(2)

(d) Find the coordinates of the point of inflexion on C.

(2)

(e) State, with a reason, whether C is concave or convex in the interval [0, 1]

(1)

(a) $y = x^3 + 6x^2 + 9x + 5$

$\frac{dy}{dx} = 3x^2 + 12x + 9$ ✔✔

$\frac{d^2y}{dx^2} = 6x + 12$ ✔

(b) When $x = -3$,

$\frac{dy}{dx} = 3(-3)^2 + 12(-3) + 9$ ✔

$= 27 - 36 + 9 = 0$

So there is a stationary point at $x = -3$. ✔

(c) When $x = -3$,

$\frac{d^2y}{dx^2} = 6(-3) + 12$ ✔

$= -18 + 12 = -6$

So when $x = -3$ there is a maximum point. ✔

(d) $6x + 12 = 0$, gives $x = -2$, $y = 3$ ✔

When $x < -2$, $6x + 12 < 0$ ⎫ change of sign; so point
When $x > -2$, $6x + 12 > 0$ ⎭ of inflexion ✔

(e) In the interval [0, 1], $6x + 12 > 0$ so C is convex. ✔

(Total for Question 4 is 10 marks)

4

95

5 A circle has centre, P, at $(-2, 7)$ and radius of 5. The tangent to the circle at T passes through the point Q (13, 12). The straight line PQ cuts the circle at W.

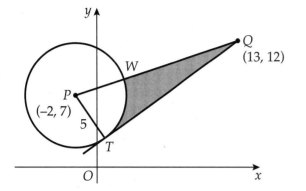

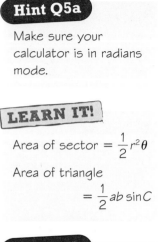

Revision Guide pages 76, 77

Hint Q5a

Make sure your calculator is in radians mode.

LEARN IT!

Area of sector $= \frac{1}{2}r^2\theta$

Area of triangle
$$= \frac{1}{2}ab\sin C$$

(a) Find the size of angle TPW in radians, giving your answer to 3 decimal places.

(4)

(b) Find the area of the shaded region TWQ, giving your answer to 2 decimal places.

(5)

Hint Q5b

Don't round any intermediate values. Use the memory functions on your calculator and write down at least 4 decimal places in your working.

(a) $PQ = \sqrt{15^2 + 5^2} = \sqrt{250}$ ✓✓

so $\cos TPW = \dfrac{5}{\sqrt{250}}$ ✓

angle $TPW = 1.249$ radians (3 d.p.) ✓

(b) Area of triangle $TPQ = \dfrac{1}{2} \times 5 \times \sqrt{250} \times \sin 1.2490...$ ✓

$= 37.5$ ✓

Area of sector $TPW = \dfrac{1}{2} \times 5^2 \times 1.2490...$ ✓

$= 15.6130...$ ✓

So shaded area $TWQ = 37.5 - 15.6130...$

$= 21.89$ (2 d.p.) ✓

(Total for Question 5 is 9 marks)

5

96

Hint

Find expressions for the vectors $\overrightarrow{AC}$, $\overrightarrow{BC}$ and $\overrightarrow{AB}$, and then find their magnitudes.

Hint

Use the cosine rule to find θ.

LEARN IT!

$$\cos A = \frac{b^2 + c^2 - a^2}{2bc}$$

6

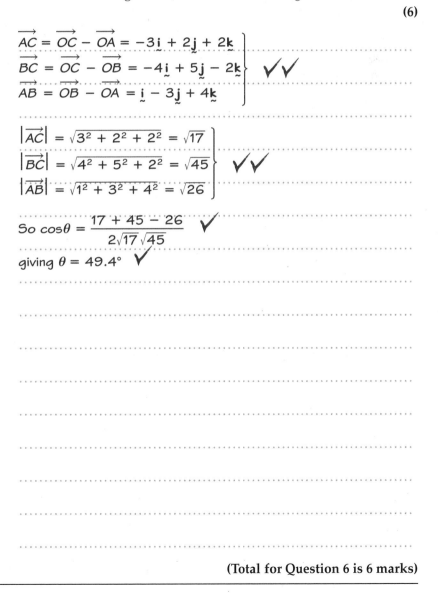

The diagram shows a sketch of triangle *ABC*. Given that, with reference to a fixed origin, *O*,

$$\overrightarrow{OA} = 2\mathbf{i} + \mathbf{j} - 3\mathbf{k}$$

$$\overrightarrow{OB} = 3\mathbf{i} - 2\mathbf{j} + \mathbf{k}$$

$$\overrightarrow{OC} = -\mathbf{i} + 3\mathbf{j} - \mathbf{k}$$

find the size of angle *ACB*, marked θ on the diagram.

(6)

$\left.\begin{array}{l}\overrightarrow{AC} = \overrightarrow{OC} - \overrightarrow{OA} = -3\mathbf{i} + 2\mathbf{j} + 2\mathbf{k} \\[4pt] \overrightarrow{BC} = \overrightarrow{OC} - \overrightarrow{OB} = -4\mathbf{i} + 5\mathbf{j} - 2\mathbf{k} \\[4pt] \overrightarrow{AB} = \overrightarrow{OB} - \overrightarrow{OA} = \mathbf{i} - 3\mathbf{j} + 4\mathbf{k}\end{array}\right\}$ ✓✓

$\left.\begin{array}{l}\left|\overrightarrow{AC}\right| = \sqrt{3^2 + 2^2 + 2^2} = \sqrt{17} \\[4pt] \left|\overrightarrow{BC}\right| = \sqrt{4^2 + 5^2 + 2^2} = \sqrt{45} \\[4pt] \left|\overrightarrow{AB}\right| = \sqrt{1^2 + 3^2 + 4^2} = \sqrt{26}\end{array}\right\}$ ✓✓

So $\cos\theta = \dfrac{17 + 45 - 26}{2\sqrt{17}\,\sqrt{45}}$ ✓

giving $\theta = 49.4°$ ✓

(Total for Question 6 is 6 marks)

6

7 $f(x) = x^3 + x + 8$

(a) Show that $f(x) = 0$ has a root α in the interval $[-2, -1]$

(2)

(b) Find $f'(x)$

(1)

(c) Taking $x_0 = -2$ as a first approximation, apply the Newton–Raphson method three times to obtain an approximate value of α. Give your answer to 3 decimal places.

(4)

Revision Guide
page 99

Hint Q7a

You need to show that $f(x)$ **changes sign** in the given interval. Make sure you state this fact in your answer.

Hint Q7c

The formula for the Newton–Raphson method is given in the formulae booklet:

$$x_{n+1} = x_n - \frac{f(x_n)}{f'(x_n)}$$

Hint Q7c

You should find that x_2 and x_3 round to the same value for α when written to 3 d.p.

(a) $f(x) = x^3 + x + 8$

$\left. \begin{array}{l} f(-2) = -8 - 2 + 8 = -2 \\ f(-1) = -1 - 1 + 8 = 6 \end{array} \right\}$ ✔ change of sign, ✔ so $-2 < \alpha - 1$ ✔

(b) $f'(x) = 3x^2 + 1$ ✔

(c) $x_0 = -2$

$x_{n+1} = x_n - \dfrac{f(x_n)}{f'(x_n)} = -2 - \left(\dfrac{-2}{13} \right)$

$x_1 = -1.846153...$ ✔

$x_2 = -1.833826...$ ✔

$x_3 = -1.833750...$ ✔

So $\alpha = -1.834$ (3 d.p.) ✔

(Total for Question 7 is 7 marks)

7

Revision Guide
page 70

Hint Q8a

The ratio between the consecutive terms is constant.

Problem solving

Use the common ratio to form an equation in k, then rearrange it. Remember that k is a positive constant.

8 The first three terms of a geometric sequence are

$$2k, (3k + 4), (9k + 7)$$

respectively, where k is a positive constant.

(a) Show that $9k^2 - 10k - 16 = 0$

(3)

(b) Find the value of k.

(2)

(c) Find the sum of the first 15 terms of this sequence, correct to 1 decimal place.

(3)

(a) $r = \dfrac{3k + 4}{2k}$ and $r = \dfrac{9k + 7}{3k + 4}$

So $\dfrac{3k + 4}{2k} = \dfrac{9k + 7}{3k + 4}$ ✓

$(3k + 4)^2 = 2k(9k + 7)$ ✓

$9k^2 + 24k + 16 = 18k^2 + 14k$

$0 = 9k^2 - 10k - 16$ ✓

(b) Solving $9k^2 - 10k - 16 = 0$:

$(9k + 8)(k - 2) = 0$ ✓

So $k = 2$ (k given as being positive). ✓

(c) $r = \dfrac{3(2) + 4}{2(2)} = \dfrac{10}{4} = 2.5$ ✓

$S_{15} = \dfrac{4(1 - 2.5^{15})}{1 - 2.5}$

$= 2483524.2$ (1 d.p.) ✓✓

(Total for Question 8 is 8 marks)

8

99

9 A curve has parametric equations

$$x = \cos\theta + \sin\theta \text{ and } y = 2\cos\theta + \sin\theta$$

Find the Cartesian equation of the curve.

(5)

$x = \cos\theta + \sin\theta$ ①

$y = 2\cos\theta + \sin\theta$ ②

② − ①: $y - x = \cos\theta$ ✓

$2 \times$ ① − ②: $2x - y = \sin\theta$ ✓

Using $\sin^2\theta + \cos^2\theta = 1$:

$$(2x - y)^2 + (y - x)^2 = 1 \text{ ✓}$$

$4x^2 - 4xy + y^2 + y^2 - 2xy + x^2 = 1$ ✓

$$5x^2 - 6xy + 2y^2 = 1 \text{ ✓}$$

Revision Guide
page 86

Hint

Solve the equations simultaneously to find expressions for $\sin\theta$ and $\cos\theta$ in terms of x and y.

Problem solving

You can sometimes use $\sin^2\theta + \cos^2\theta = 1$ to eliminate the parameter in parametric equations.

Hint

Your final answer can be given implicitly. You don't need to be able to write it in the form $y = \ldots$

(Total for Question 9 is 5 marks)

9

Hint Q10a

Differentiate every term with respect to x, then collect all terms involving $\dfrac{dy}{dx}$ on one side, and the remaining terms on the other side.

LEARN IT!

Use the product rule, or learn the rule for differentiating xy terms implicitly:

$$\frac{d}{dx}(xy) = x\frac{dy}{dx} + y$$

Problem solving

The tangent to C will be parallel to the y-axis, which means the gradient is infinite.

10 The curve C has equation

$$x^2 + xy = 2y^2 + 63$$

(a) Find $\dfrac{dy}{dx}$ in terms of x and y.

(5)

(b) A point P lies on C. The tangent to C at the point P is parallel to the y-axis.

Find the possible coordinates of P.

(5)

(a) $x^2 + xy = 2y^2 + 63$

Differentiating implicitly:

$$2x + \left(x\frac{dy}{dx} + 1(y)\right) = 4y\frac{dy}{dx} \quad \checkmark\checkmark\checkmark$$

$$2x + y = (4y - x)\frac{dy}{dx} \quad \checkmark$$

So $\qquad \dfrac{dy}{dx} = \dfrac{2x + y}{4y - x} \quad \checkmark$

(b) At P, the gradient is infinite, so $4y - x = 0$

So, at P, $x = 4y \quad \checkmark$

Substituting for x in the equation for C:

$$16y^2 + 4y^2 = 2y^2 + 63 \quad \checkmark$$

$$18y^2 = 63$$

$$y = \pm\sqrt{\frac{63}{18}} = \pm\sqrt{3.5} = \pm1.87082\ldots$$

This gives $y = \pm1.87$ (3 s.f.) $\quad \checkmark$

So the possible coordinates of P are

$(-7.48, -1.87) \quad \checkmark$ and $(7.48, 1.87) \quad \checkmark$

(Total for Question 10 is 10 marks)

10

101

11 The point P is the point on the curve

$$x = 3 \cot\left(\frac{\pi}{2} - y\right)$$

with y-coordinate $\frac{\pi}{6}$

Find the equation of the normal to the curve at P.

(8)

Revision Guide pages 38, 92

Hint

The normal to the curve at P is perpendicular to the tangent at P.

Hint

Use $y = \frac{\pi}{6}$ to find the x-coordinate of P.

Problem solving

You are given x in terms of y, so first find $\frac{dx}{dy}$, then invert to find $\frac{dy}{dx}$.

$x = 3 \cot\left(\frac{\pi}{2} - y\right)$

When $y = \frac{\pi}{6}$, $x = 3 \cot\left(\frac{\pi}{2} - \frac{\pi}{6}\right)$ ✓

$\qquad = 3 \cot\frac{\pi}{3}$

$\qquad = \frac{3}{\tan\frac{\pi}{3}} = \frac{3}{\sqrt{3}} = \sqrt{3}$ ✓

$\frac{dx}{dy} = 3\left(-\text{cosec}^2\left(\frac{\pi}{2} - y\right)\right)(-1)$ ✓

$\qquad = 3\,\text{cosec}^2\left(\frac{\pi}{2} - y\right)$ ✓

$\frac{dy}{dx} = \frac{1}{3}\sin^2\left(\frac{\pi}{2} - y\right)$ ✓

When $y = \frac{\pi}{6}$, $\frac{dy}{dx} = \frac{1}{3}\sin^2\frac{\pi}{3}$ ✓

$\qquad = \frac{1}{3}\left(\frac{\sqrt{3}}{2}\right)^2 = \frac{1}{4}$ ✓

Gradient of normal at $\left(\sqrt{3}, \frac{\pi}{6}\right) = -4$

Equation of normal is $y - \frac{\pi}{6} = -4(x - \sqrt{3})$

i.e. $\qquad\qquad 4x + y = \frac{\pi}{6} + 4\sqrt{3}$ ✓

(Total for Question 11 is 8 marks)

11

12 (a) Prove that $\operatorname{cosec} 2A - \cot 2A \equiv \tan A$

(4)

   (b) Hence, or otherwise, solve for $0° < \theta < 180°$,

$$\operatorname{cosec}(4\theta - 20°) - \cot(4\theta - 20°) = \sqrt{3}$$

(4)

Hint Q12a

Rewrite $\operatorname{cosec} 2A$ and $\cot 2A$ in terms of $\sin 2A$ and $\cos 2A$, Then use the double angle formulae.

(a) $\operatorname{cosec} 2A - \cot 2A = \dfrac{1}{\sin 2A} - \dfrac{\cos 2A}{\sin 2A}$

$$= \dfrac{1 - \cos 2A}{\sin 2A} \checkmark$$

$$= \dfrac{1 - (1 - 2\sin^2 A)}{2\sin A \cos A} \checkmark$$

$$= \dfrac{2\sin^2 A}{2\sin A \cos A} \checkmark$$

$$= \dfrac{\sin A}{\cos A}$$

$$= \tan A \checkmark$$

Hint Q12b

Make sure you have found all the possible answers in the given range – there should be two.

(b) $\operatorname{cosec}(4\theta - 20°) - \cot(4\theta - 20°) = \sqrt{3}$

now becomes $\tan(2\theta - 10°) = \sqrt{3}$ $\checkmark$

If $0° < \theta < 180°$, then $-10° < (2\theta - 10°) < 350°$

Principal value $= \tan^{-1}\sqrt{3} = 60°$ $\checkmark$

So $2\theta - 10° = 60°$, $\theta = 35°$ $\checkmark$

or $2\theta - 10° = 240°$, $\theta = 125°$ $\checkmark$

(Total for Question 12 is 8 marks)

12

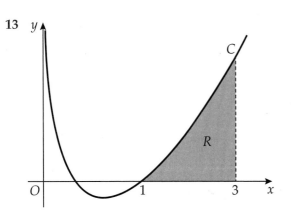

13

The diagram shows a sketch of part of the curve, C, with equation

$$y = (3x - 1) \ln x, \ x > 0$$

The finite region R is bounded by the curve C, the x-axis and the line with equation $x = 3$.

The table shows corresponding values of x and y, given to 4 decimal places, as appropriate.

x	1	1.5	2	2.5	3
y	0	1.4191	3.4657	5.9559	8.7889

(a) Use the trapezium rule, with all the values of y in the table, to obtain an estimate for the area of R, giving your answer to 3 decimal places.

(3)

(b) Explain how the trapezium rule could be used to obtain a more accurate estimate for the area of R.

(1)

(c) Show that the exact area of R can be written in the form $\dfrac{a}{b} \ln c + d$,

where a, b, c and d are integers to be found.

(7)

(In part (c), solutions based entirely on graphical or numerical methods are not acceptable.)

(a) Area of R

$$\approx \frac{0.5}{2} \checkmark \ [0 + 8.7889 + 2(1.4191 + 3.4657 + 5.9559)] \checkmark$$

$$= 7.617575 = 7.618 \ (3 \ \text{d.p.}) \checkmark$$

(b) Any valid reason, e.g. increase the number of strips, decrease the width of strips. $\checkmark$

Revision Guide
pages 104, 108

Hint Q13a

There are five values given so you need to use the trapezium rule with four strips.

Hint Q13b

The question says 'explain' so give an answer in words.

Hint Q13c

Remember to set $\ln x$ as u when using integration by parts.

Problem solving

You will need to use integration by parts twice in part (c).

Watch out!

You need to give your answer to part (c) in the correct form. You could write out the values of a, b, c and d to make sure.

13

(c) Area $= \int_1^3 3x \ln x \, dx - \int_1^3 \ln x \, dx$ ✔

For $\int_1^3 3x \ln x \, dx$, put $\dfrac{dv}{dx} = 3x$ and $u = \ln x$

Integral $= \left[\dfrac{3x^2}{2}\ln x\right]_1^3 - \int_1^3 \dfrac{3x^2}{2}\left(\dfrac{1}{x}\right)dx$ ✔

$= \left[\dfrac{3x^2}{2}\ln x - \dfrac{3x^2}{4}\right]_1^3$ ✔

For $\int_1^3 \ln x \, dx = \int_1^3 1 \cdot (\ln x)\, dx$, put $\dfrac{dv}{dx} = 1$ and $u = \ln x$

Integral $= [x\ln x]_1^3 - \int_1^3 x\left(\dfrac{1}{x}\right)dx$ ✔

$= [x\ln x - x]_1^3$ ✔

So,

area of $R = \left[\dfrac{3x^2}{2}\ln x - \dfrac{3x^2}{4} - x\ln x + x\right]_1^3$

$= \left(\dfrac{27}{2}\ln 3 - \dfrac{27}{4} - 3\ln 3 + 3\right) - \left(0 - \dfrac{3}{4} - 0 + 1\right)$ ✔

$= \dfrac{21}{2}\ln 3 - 4$ ✔

(which is 7.535429..., so 7.618 was a good approximation).

(Total for Question 13 is 11 marks)

TOTAL FOR PAPER IS 100 MARKS

Paper 2: Pure Mathematics 2

Answer all questions. Write your answers in the spaces provided.

1 The curve C has equation $y = \text{f}(x)$ where

$$\text{f}(x) = \frac{5x - 1}{x + 3}, \ x \neq -3$$

(a) Show that $\text{f}'(x) = \dfrac{16}{(x + 3)^2}$

(3)

Hint Q1a

Use the quotient rule
$u = 5x - 1$ and
$v = x + 3$

The diagram shows the graph of $y = \text{f}(x)$. The point P with x-coordinate 1 lies on C.

The line L_1 is a tangent to C at P. The line L_2 is also a tangent to C at Q, and is parallel to L_1.

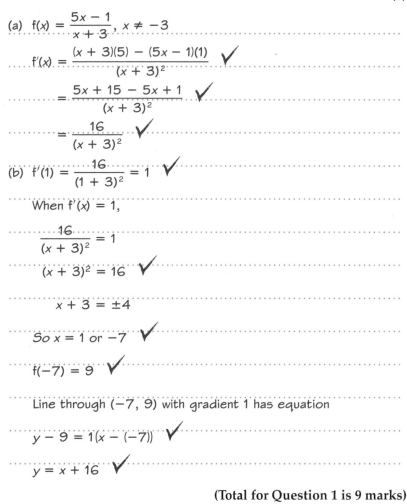

(b) Find the equation of L_2.

(6)

LEARN IT!

You can learn the quotient rule for a function $\dfrac{u}{v}$ as $\dfrac{u'v - v'u}{v^2}$

Watch out!

Make sure you give the whole equation of L_2.

(a) $\text{f}(x) = \dfrac{5x - 1}{x + 3}, \ x \neq -3$

$\text{f}'(x) = \dfrac{(x + 3)(5) - (5x - 1)(1)}{(x + 3)^2}$ ✓

$= \dfrac{5x + 15 - 5x + 1}{(x + 3)^2}$ ✓

$= \dfrac{16}{(x + 3)^2}$ ✓

(b) $\text{f}'(1) = \dfrac{16}{(1 + 3)^2} = 1$ ✓

When $\text{f}'(x) = 1$,

$\dfrac{16}{(x + 3)^2} = 1$

$(x + 3)^2 = 16$ ✓

$x + 3 = \pm 4$

So $x = 1$ or -7 ✓

$\text{f}(-7) = 9$ ✓

Line through $(-7, 9)$ with gradient 1 has equation

$y - 9 = 1(x - (-7))$ ✓

$y = x + 16$ ✓

(Total for Question 1 is 9 marks)

15

Revision Guide
page 65

Hint Q2a

First sketch $y = |f(x)|$ then reflect in the x-axis.

LEARN IT!

A graph in the form $y = |f(x)|$ will be entirely above the x-axis, so a graph in the form $y = -|f(x)|$ will be entirely below the x-axis.

Hint Q2b

There are three stages: reflect in the y-axis, then apply scale factor of $\frac{1}{2}$, and finally a translation.

2

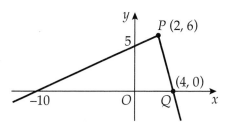

The diagram shows part of a graph with equation $y = f(x)$, $x \in \mathbb{R}$.

The graph consists of two line segments that meet at the point $P\,(2, 6)$. The graph also crosses the axes at $Q\,(4, 0)$ and two other points, $(-10, 0)$ and $(0, 5)$.

Sketch, on separate diagrams, the graphs of

(a) $y = -|f(x)|$

(2)

(b) $y = \frac{1}{2}f(-x) - 2$

(2)

Label the points P' and Q' on your sketch graphs.

(c) Does $f^{-1}(x)$ exist? Explain your answer.

(1)

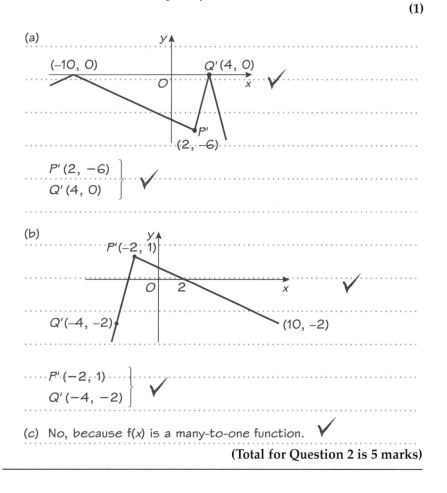

(a)

$P'\,(2, -6)$
$Q'\,(4, 0)$ ✓

(b)

$P'\,(-2, 1)$
$Q'\,(-4, -2)$ ✓

(c) No, because $f(x)$ is a many-to-one function. ✓

(Total for Question 2 is 5 marks)

16

3 $g(x) = x^3 + ax^2 - ax + 80$, where a is a constant.

Given that $g(-8) = 0$

(a) (i) show that $a = 6$

     (ii) express $g(x)$ as a product of two algebraic factors.

     (4)

Given that $2\log_5(x + 3) + \log_5 x - \log_5(3x - 16) = 1$

(b) show that $x^3 + 6x^2 - 6x + 80 = 0$

     (4)

(c) Hence explain why

$$2\log_5(x + 3) + \log_5 x - \log_5(3x - 16) = 1$$

has no real roots.

     (2)

(a) (i) $g(-8) = 0$

$0 = (-8)^3 + a(-8)^2 - a(-8) + 80$ ✓

$0 = -512 + 64a + 8a + 80$

$432 = 72a$

$6 = a$ ✓

(ii) $g(x) = x^3 + 6x^2 - 6x + 80$

$= (x + 8)(x^2 - 2x + 10)$ ✓✓

(b) $2\log_5(x + 3) + \log_5 x - \log_5(3x - 16) = 1$

$\log_5(x + 3)^2 + \log_5 x - \log_5(3x - 16) = 1$ ✓

$\log_5 \dfrac{x(x + 3)^2}{3x - 16} = 1$ ✓

$\dfrac{x(x + 3)^2}{3x - 16} = 5$ ✓

$x(x + 3)^2 = 15x - 80$

$x^3 + 6x^2 + 9x - 15x + 80 = 0$

$x^3 + 6x^2 - 6x + 80 = 0$ ✓

(c) $g(x) = 0$ has only one solution, $x = -8$

because $x^2 - 2x + 10 = 0$ has no real solutions ✓

When $x = -8$, $\log_5(x + 3)$, $\log_5 x$ and $\log_5(3x - 16)$ would all

be logarithms of negative numbers, which are not defined.

So $2\log_5(x + 3) + \log_5 x - \log_5(3x - 16) = 1$ has no real

roots. ✓

(Total for Question 3 is 10 marks)

18

4

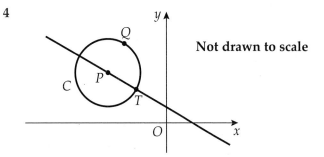

Not drawn to scale

Revision Guide
pages 20, 21, 22

The circle C has centre P.

Points Q and T lie on C.

The equation of the line PT is $2x + 3y = 6$

The line through P perpendicular to PT passes through the point Q $(-5, 14)$.

(a) Show that an equation of the line PQ is

$$3x - 2y = -43$$

(3)

(b) Find an equation for C.

(4)

The line with equation $3x - 2y = k$ is a tangent to C at the point T.

(c) Find the value of the constant k.

(3)

Hint Q4a

Rearrange the equation of PT into the form $y = mx + c$ to find the gradient.

Then use $\dfrac{-1}{m}$ to find the perpendicular gradient.

Hint Q4b

Find the coordinates of P and use radius = distance PQ.

Hint Q4c

Work out the coordinates of T.

(a) Gradient of $PT = \dfrac{-2}{3}$, so gradient of $PQ = \dfrac{3}{2}$ ✓

Equation of PQ: $y - 14 = \dfrac{3}{2}(x + 5)$ ✓

$2y - 28 = 3x + 15$

$-43 = 3x - 2y$ ✓

19

110

(b) Solve $2x + 3y = 6$ and $3x - 2y = -43$ simultaneously to find coordinates of P: ✓

$$2x + 3y = 6 \qquad ①$$

$$3x - 2y = -43 \qquad ②$$

① × 2: $4x + 6y = 12$

② × 3: $9x - 6y = -129$

$$13x = -117$$

$$x = -9$$

$$y = 8$$

So the point P has coordinates $(-9, 8)$ ✓

Radius of circle = distance PQ

$$= \sqrt{(-9 - (-5))^2 + (8 - 14)^2}$$

$$= \sqrt{4^2 + 6^2}$$

$$= \sqrt{52} \quad ✓$$

Equation of C is $(x + 9)^2 + (y - 8)^2 = 52$ ✓

(c)

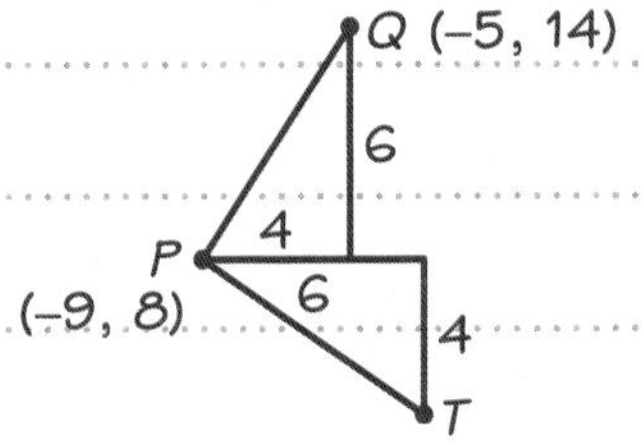

Coordinates of T are $(-3, 4)$ ✓

Substitute $(-3, 4)$ in $3x - 2y = k$: ✓

$$3(-3) - 2(4) = k$$

$$k = -17 \quad ✓$$

(Total for Question 4 is 10 marks)

20

111

5   The curve C has equation $y = 2x^3 - 3x$

The point P (2, 10) lies on C.

Use differentiation from first principles to find the value of the gradient of the tangent to C at P.

(5)

$y = 2x^3 - 3x$

$f'(x) = \lim_{h \to 0} \dfrac{f(x + h) - f(x)}{h}$

$= \lim_{h \to 0} \dfrac{[2(x + h)^3 - 3(x + h)] - (2x^3 - 3x)}{h}$ ✓

$= \lim_{h \to 0} \dfrac{2x^3 + 6x^2h + 6xh^2 + 2h^3 - 3x - 3h - 2x^3 + 3x}{h}$ ✓

$= \lim_{h \to 0} \dfrac{6x^2h + 6xh^2 + 2h^3 - 3h}{h}$

$= \lim_{h \to 0} (6x^2 + 6xh + 2h^2 - 3)$ ✓

As $h \to 0$, $6xh \to 0$ and $2h^2 \to 0$ ✓

So $f'(x) = 6x^2 - 3$, and at P, $f'(x) = 6(2)^2 - 3 = 21$ ✓

(Total for Question 5 is 5 marks)

Revision Guide
page 35

LEARN IT!

The rule for differentiating from first principles is given in the formulae booklet:

$f'(x) = \lim_{h \to 0} \dfrac{f(x + h) - f(x)}{h}$

Hint

Take care when expanding brackets and use correct mathematical language and notation.

Problem solving

Don't round any intermediate values. Write down at least 4 decimal places or use the memory functions on your calculator.

Hint Q6b

Check that all your final solutions are within the specified range.

Problem solving

You can use a sketch graph to check that you have found all the possible solutions.

6 (a) Write $6 \cos \theta + 11 \sin \theta$ in the form

$R \cos(\theta - \alpha)$ where $R \geqslant 0$ and $0 < \alpha < \dfrac{\pi}{2}$

(3)

(b) Hence, solve the equation

$6 \cos \theta + 11 \sin \theta = 8$

for $0 \leqslant \theta \leqslant 2\pi$, giving your answers to 2 decimal places.

(5)

(a) Write $6 \cos\theta + 11 \sin\theta \neq R\cos\theta \cos\alpha + R\sin\theta \sin\alpha$

So, $R \cos\alpha = 6$, $R \sin\alpha = 11$

$\tan\alpha = \dfrac{11}{6}$, so $\alpha = 1.071449...$ ✔

$R^2 = 6^2 + 11^2 = 157$, so $R = \sqrt{157}$ ✔

So $6 \cos\theta + 11 \sin\theta = \sqrt{157}\cos(\theta - 1.071449...)$ ✔

(b) $\sqrt{157}\cos(\theta - 1.071449...) = 8$

$\cos(\theta - 1.071449...) = \dfrac{8}{\sqrt{157}}$ ✔

If $0 \leqslant \theta \leqslant 2\pi$, range of values becomes

$-1.071449...$ to $5.211735...$ rad

and principal value of

$\cos^{-1}\left(\dfrac{8}{\sqrt{157}}\right)$ is $0.878288...$ rad

So $\theta - 1.071449... = 0.878288$ ✔

giving $\theta = 1.949737... = 1.95$ radians (2 d.p.) ✔

or $\theta - 1.071449... = -0.878288$ ✔

giving $\theta = 0.193161... = 0.19$ radians (2 d.p.) ✔

(Total for Question 6 is 8 marks)

7 A hot metal sphere is cooled by placing it into a liquid. The temperature, $T°C$, after t minutes in the liquid, is given by

$T = 340\,e^{-0.04t} + 23, t \geqslant 0$

(a) Find the temperature of the sphere as it enters the liquid.

(1)

(b) Find the value of t when $T = 300$, giving your answer to 3 significant figures.

(2)

(c) Find the rate at which the temperature of the sphere is decreasing at the instant when $t = 45$. Give your answer in $°C$ per minute, to 3 significant figures.

(3)

(d) Sketch a graph of T against t, showing clearly any points where the graph crosses or touches the axes, and any asymptotes.

(2)

3 hours after the metal sphere was put into the liquid to cool, its temperature was measured as $21.6\,°C$.

(e) Using this information, evaluate the model, explaining your reasoning.

(1)

(a) $T = 340\,e^{-0.04t} + 23, t \geqslant 0$

When $t = 0$, $T = 340 \times 1 + 23 = 363$ ✔

(b) When $T = 300$, $300 = 340\,e^{-0.04t} + 23$ ✔

$$277 = 340\,e^{-0.04t}$$

$$e^{-0.04t} = \frac{277}{340}$$

$$-0.04t = \ln\frac{277}{340}$$

So $t = 5.12$ (3 s.f.) ✔

(c) $\frac{dT}{dt} = (-340 \times 0.04)\,e^{-0.04t}$ ✔

So, when $t = 45$, $\frac{dT}{dt} = (-340 \times 0.04)\,e^{-0.04 \times 45}$ ✔

$$= -2.25°C \text{ per minute (3 s.f.)} ✔$$

Revision Guide
page 52

Hint Q7a

Substitute $t = 0$ into the equation.

Hint Q7c

The rate of change of the temperature will be given by $\frac{dT}{dt}$. This value will be negative, because the temperature is falling.

Watch out!

Sketch graphs should still be drawn neatly. Make sure you label the axes, the origin, any asymptotes, and any points of intersection with the axes.

Modelling

Compare the recorded value with the value predicted by the model.

(d)

(e) The model predicts that the temperature will not fall below 23°, so this measurement suggests the model is not suitable over this range of values of t. ✓

(Total for Question 7 is 9 marks)

24

8 $\dfrac{29 - 6x - 2x^2}{(x + 4)(3 - x)} \equiv A + \dfrac{B}{x + 4} + \dfrac{C}{3 - x}$

Revision Guide
pages 39, 58

(a) Find the values of the constants A, B and C.

(4)

Hint Q8a

One approach for this question is to multiply both sides by $(x + 4)(3 - x)$ then compare coefficients.

$$f(x) = \dfrac{29 - 6x - 2x^2}{(x + 4)(3 - x)},\ x > 3$$

(b) Prove that $f(x)$ is a decreasing function.

(3)

Hint Q8b

Differentiate the three terms on the right-hand side of the original identity. You need to show that the expression for $f'(x) \leq 0$ for all values of $x > 3$.

(a) $29 - 6x - 2x^2 \equiv A(x + 4)(3 - x) + B(3 - x) + C(x + 4)$ ✓

Equating coefficients of x^2:

$-2 = -A$, so $A = 2$ ✓

Let $x = 3$:

$29 - 18 - 18 = 7C$

$-7 = 7C$

$C = -1$ ✓

Equating constant terms:

$29 = 12A + 3B + 4C$

$29 = 24 + 3B - 4$

$9 = 3B$

$B = 3$ ✓

(b) $f(x) = 2 + 3(x + 4)^{-1} - 1(3 - x)^{-1}$

$f'(x) = -3(x + 4)^{-2}(1) + 1(3 - x)^{-2}(-1)$ ✓

$= \dfrac{-3}{(x + 4)^2} - \dfrac{1}{(3 - x)^2}$ ✓

Since $(x + 4)^2$ and $(3 - x)^2$ are always > 0 for $x > 3$, $f'(x)$ is

always negative; hence $f(x)$ is a decreasing function. ✓

(Total for Question 8 is 7 marks)

25

Problem solving

You will need to choose a suitable substitution in the form $u = f(x)$. Look for substitutions where you can express part of the integrand easily in terms of u, and part of it in terms of $\frac{du}{dx}$

Hint

Using your substitution, obtain an integral in u and du only.

Watch out!

Make sure you convert the limits from x-values to u-values.

9 Using a suitable substitution, or otherwise, use an algebraic method to find the exact value of

$$\int_{\frac{\pi}{12}}^{\frac{\pi}{4}} \sin^4 2x \cos 2x \, dx$$

(6)

$$\int_{\frac{\pi}{12}}^{\frac{\pi}{4}} \sin^4 2x \cos 2x \, dx$$

$u = \sin 2x$ ✓

$\frac{du}{dx} = 2\cos 2x$

$du = 2\cos 2x \, dx$ ✓

When $x = \frac{\pi}{12}$, $u = \sin\frac{\pi}{6} = 0.5$

When $x = \frac{\pi}{4}$, $u = \sin\frac{\pi}{2} = 1$ ✓

So, $\int_{\frac{\pi}{12}}^{\frac{\pi}{4}} \sin^4 2x \cos 2x \, dx = \int_{0.5}^{1} \frac{u^4}{2} du$ ✓

$= \left[\frac{u^5}{10}\right]_{0.5}^{1}$ ✓

$= \frac{1}{10} - \frac{1}{320}$

$= \frac{31}{320}$

$= 0.096875$ ✓

(Total for Question 9 is 6 marks)

10 Solve, for $-180° \leq \theta \leq 180°$

$$2 \sin \theta \tan \theta = \sin \theta + 6 \cos \theta$$

(9)

Revision Guide
page 79

$$2 \sin \theta \tan \theta = \sin \theta + 6 \cos \theta$$

$$2 \sin \theta \frac{\sin \theta}{\cos \theta} = \sin \theta + 6 \cos \theta \quad ✓$$

$$2 \sin^2 \theta = \sin \theta \cos \theta + 6 \cos^2 \theta$$

$$2 \sin^2 \theta - \sin \theta \cos \theta - 6 \cos^2 \theta = 0 \quad ✓$$

$$(2 \sin \theta + 3 \cos \theta)(\sin \theta - 2 \cos \theta) = 0 \quad ✓$$

So, $2 \sin \theta + 3 \cos \theta = 0$

$$2 \sin \theta = -3 \cos \theta$$

$$\tan \theta = \frac{-3}{2} \quad ✓$$

$\theta = -56.3°$ ✓ or $123.7°$ ✓ (1 d.p.)

or $\sin \theta - 2 \cos \theta = 0$

$$\sin \theta = 2 \cos \theta$$

$$\tan \theta = 2 \quad ✓$$

$\theta = -116.6°$ ✓ or $63.4°$ ✓ (1 d.p.)

Problem solving

When equations have terms in sin, cos and tan, it is usually better to express the equation in terms of sin and cos only.

Hint

Once you have written the equation in terms of sin and cos, you can factorise it.

(Total for Question 10 is 9 marks)

Revision Guide page 75

Problem solving

Take out a factor of $9^{\frac{1}{2}}$ before using the binomial expansion.

Keep the $\left(\dfrac{4x}{9}\right)$ term in brackets so that you don't make any mistakes when squaring.

Watch out!

Remember to fully simplify all coefficients.

11 (a) Find the binomial expansion of

$$(9 + 4x)^{\frac{1}{2}}, \ |x| < \frac{9}{4}$$

in ascending powers of x, up to and including the term in x^2. Give each coefficient in its simplest form.

(5)

(b) Find the exact value of $(9 + 4x)^{\frac{1}{2}}$ when $x = 0.75$. Give your answer in the form $k\sqrt{3}$, where k is a constant to be found.

(2)

(c) Substitute $x = 0.75$ into your binomial expansion and hence find an approximate value for $\sqrt{3}$. Give your answer to 3 decimal places.

(3)

(a) $(9 + 4x)^{\frac{1}{2}} = \left[9\left(1 + \dfrac{4x}{9}\right)\right]^{\frac{1}{2}}$ ✓

$= 3\left(1 + \dfrac{4x}{9}\right)^{\frac{1}{2}}$ ✓

$= 3\left[1 + \dfrac{1}{2}\left(\dfrac{4x}{9}\right) + \dfrac{\left(\frac{1}{2}\right)\left(-\frac{1}{2}\right)}{1 \times 2}\left(\dfrac{4x}{9}\right)^2 + \dots\right]$ ✓

$= 3\left(1 + \dfrac{2x}{9} - \dfrac{2}{81}x^2 + \dots\right)$ ✓

$\approx 3 + \dfrac{2x}{3} - \dfrac{2}{27}x^2$ ✓

(b) When $x = 0.75$:

$(9 + 4x)^{\frac{1}{2}} = (9 + 3)^{\frac{1}{2}} = 12^{\frac{1}{2}}$ ✓

$= \sqrt{12} = 2\sqrt{3}$ ✓

(c) When $x = 0.75$:

binomial expansion $= 3 + 0.5 - \dfrac{2}{27}(0.75)^2 + \dots$ ✓

$\approx 3 + 0.5 - 0.041666\dots$

$= 3.458333\dots$ ✓

So, $2\sqrt{3} \approx 3.458333\dots$

$\sqrt{3} \approx 1.729166\dots$

$= 1.729$ (3 d.p.) ✓

(Total for Question 11 is 10 marks)

28

12

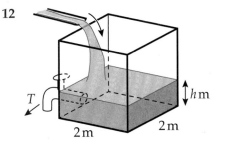

Revision Guide
pages 41, 97, 109

The diagram shows a water tank with a square base of side 2 m. Water is flowing into the tank at a constant rate of 0.5 m³ per minute.

At time t minutes, the depth of the water in the tank is h metres.

When a tap, T, is opened at the bottom of the tank, water leaves the tank at a rate of $0.4h$ m³ per minute.

(a) Show that t minutes after the tap has opened,

$$40\frac{dh}{dt} = 5 - 4h$$

(5)

When $t = 0$, $h = 0.6$

(b) Solve the differential equation from part (a) to show that

$$t = A\ln\left(\frac{B}{5 - 4h}\right)$$

where A and B are constants to be found.

(5)

(c) Find the value of t when $h = 0.9$. Give your answer to 2 decimal places.

(2)

Modelling

Use the connection between $\dfrac{dV}{dt}$ and $\dfrac{dh}{dt}$ to set up an equation for $\dfrac{dh}{dt}$, the rate of change of height with respect to time.

Problem solving

Use separation of variables to solve the differential equation. Remember to include a constant of integration and evaluate it using the initial conditions given.

(a) Volume $V = 2 \times 2 \times h = 4h$

$$\frac{dV}{dh} = 4 \quad ✓$$

Rate of change of volume, $\dfrac{dV}{dt} = \dfrac{dV}{dh} \times \dfrac{dh}{dt}$ ✓

i.e. $0.5 - 0.4h$ ✓ $= 4 \times \dfrac{dh}{dt}$ ✓

$$5 - 4h = 40\frac{dh}{dt} \quad ✓$$

(b) Rearranging and integrating using separation of variables:

$$\int \frac{1}{40} dt = \int \frac{1}{5 - 4h} dh \quad \checkmark$$

$$\frac{t}{40} = -\frac{1}{4} \ln(5 - 4h) + c \quad \checkmark$$

$$t = -10 \ln(5 - 4h) + c'$$

When $t = 0$, $h = 0.6$

so $\quad 0 = -10 \ln(2.6) + c'$

$10 \ln(2.6) = c' \quad \checkmark$

So, $t = -10\ln(5 - 4h) + 10\ln(2.6) \quad \checkmark$

$$t = 10 \ln\left(\frac{2.6}{5 - 4h}\right) \quad \checkmark$$

(c) When $h = 0.9$, $t = 10\ln\left(\frac{2.6}{1.4}\right) \quad \checkmark$

$$= 6.19 \text{ minutes (2 d.p.)} \quad \checkmark$$

(Total for Question 12 is 12 marks)

TOTAL FOR PAPER IS 100 MARKS

Paper 3: Statistics and Mechanics
SECTION A: STATISTICS

Answer all questions. Write your answers in the spaces provided.

1 'Super Spicy' chilli powder is sold in packets. A shopkeeper measures the masses of the contents of a random sample of 80 packets of 'Super Spicy' chilli powder from his stock. The results are shown in the table.

Mass, w (g)	Midpoint, x (g)	Frequency, f
$391 \leqslant w < 395$	393	6
$395 \leqslant w < 398$	396.5	12
$398 \leqslant w < 402$	400	34
$402 \leqslant w < 407$	404.5	18
$407 \leqslant w < 410$	408.5	10

(You may use $\Sigma fx^2 = 12\,867\,128$)

A histogram is drawn and the class $395 \leqslant w < 398$ is represented by a rectangle of width 1.5 cm and height 6 cm.

(a) Calculate the width and height of the rectangle representing the class $402 \leqslant w < 407$

(3)

(b) Use linear interpolation to estimate the median mass of the contents of a packet of 'Super Spicy' chilli powder.

(2)

(c) Estimate the mean and the standard deviation of the mass of the contents of a packet of 'Super Spicy' chilli powder to 1 decimal place.

(3)

The shopkeeper claims that the mean mass of the contents of the packets is more than the stated mass. Given that the stated mass of a packet of 'Super Spicy' chilli powder is 400 g and that the actual standard deviation is 4 g:

(d) test, using a 2% level of significance, whether or not the shopkeeper's claim is justified. State your hypotheses clearly. (You may assume that the mass of the contents of a packet is normally distributed.)

(5)

(e) Using your answers to parts (b) and (c), comment on the assumption that the mass of the contents of a packet is normally distributed.

(1)

31

(a) For $395 \leqslant w < 398$, area = 9 and frequency = 12

So area : frequency = 3 : 4 ✓

For $402 \leqslant w < 407$, the width will be $\frac{5}{3} \times 1.5 = 2.5\,\text{cm}$ ✓

Frequency = 18, so area = $\frac{3}{4} \times 18 = 13.5$

and height = $\frac{13.5}{2.5} = 5.4\,\text{cm}$ ✓

(b) Median is in the $398 \leqslant w < 402$ class

Median = $398 + \left(\frac{22}{34} \times 4\right)$ ✓

$= 400.588...$

$= 400.6\,\text{g}$ (1 d.p.) ✓

(c) Using a calculator, $\bar{x} = 401.025... = 401.0\,\text{g}$ (1 d.p.) ✓

$\sigma = 4.2484... = 4.2\,\text{g}$ (1 d.p.) ✓✓

(d) Sample size = 80, so use $\text{N}\left(400, \left(\frac{4}{\sqrt{80}}\right)^2\right)$ ✓

$= \text{N}(400, 0.4472^2)$

$H_0: \mu = 400, \quad H_1: \mu > 400$ ✓

$P(X > 401) = 1 - 0.9873$

$= 0.0127$ ✓

Testing at the 2% level, $0.0127 < 0.02$ ✓

so reject H_0 and accept H_1: the shopkeeper's claim that the mean mass is greater than the stated mass of 400 g is justified. ✓

(e) From parts (b) and (c), median = 400.6 g and mean = 401 g

These are very close, so the assumption that the mass is normally distributed is very reasonable. ✓

(Total for Question 1 is 14 marks)

2 The table shows the mean daily temperature ($t°$C) and the mean daily rainfall (h mm) for the month of June 2015, for the seven places in the northern hemisphere from the large data set.

	A	B	C	D	E	F	G
$t°$C	12.8	13.3	16.8	15.1	13.8	24.7	26.4
h mm	1.52	0.81	0.58	1.10	1.67	1.60	8.60

(a) Calculate the product moment correlation coefficient for these data.

(1)

(b) Stating your hypotheses clearly, test, at the 5% level of significance, whether or not the product moment correlation coefficient is greater than 0.

(3)

(c) Using your knowledge of the large data set, suggest the names of the places labelled **F** and **G** in the table.

(1)

(d) Suggest how you could make better use of the large data set to investigate the relationship between the mean daily temperature and the mean daily rainfall, justifying your answer.

(2)

Revision Guide
page 137

Hint Q2a

Use your calculator to find the PMCC.

Hint Q2b

Write null and alternative hypotheses in terms of the population PMCC, p.

Hint Q2b

Compare your value of the product moment correlation coefficient with the 5% tabulated value for $n = 7$.

Remember to state your conclusion in the context of the original data.

(a) $r = 0.7058$ ✓

(b) H_0: $\rho = 0$, H_1: $\rho > 0$ ✓

  $n = 7$, 5% value = 0.6694, critical region is $r > 0.6694$

  Since 0.7058 > 0.6694, there is sufficient evidence to

  reject H_0 and accept H_1 ✓

  So there does appear to be a linear relationship between

  the mean daily temperature and the mean daily rainfall for this

  month. ✓

(c) Beijing and Jacksonville ✓

(d) Consider data from additional months or years. ✓

  Data for one month can be atypical, so include July and August, in

  case their data show different patterns. ✓

(Total for Question 2 is 7 marks)

33

Revision Guide
pages 140, 142, 145

Hint Q3a

Use your calculator with the values given in the question. Write down your answer to four decimal places.

Hint Q3c

Use the inverse normal function on your calculator.

Hint Q3d

Remember to work out the variance for the sample, $\frac{\sigma^2}{n}$, before you use a hypothesis test.

3   A firm manufactures energy-saving halogen light bulbs. The random variable, X, represents the lifetime, in hours, of a particular light bulb.

X is normally distributed with mean 2000 hours and standard deviation 150 hours.

(a) Find $P(1900 < X < 2200)$

(1)

(b) The manufacturer claims that 95% of these light bulbs last longer than 1750 hours. Is this claim valid?

(2)

(c) Find the time, h hours, such that 99% of light bulbs last at least h hours.

(2)

The manufacturer makes a slight change to the process for producing the light bulbs. They now claim an increase in the mean lifetime of the light bulbs and a standard deviation of 120 hours.

(d) A sample of 30 light bulbs was chosen and found to have a mean lifetime of 2040 hours.

Stating your hypotheses clearly, and using a 1% level of significance, test whether or not these findings support the claim that the mean lifetime with the new process is more than 2000 hours.

(5)

(a) $P(1900 < X < 2200) = 0.6563$ ✓

(b) $P(X > 1750) = 0.9522$ ✓

This is greater than 0.95, so the claim is valid. ✓

(c) Find h such that $P(X > h) \geqslant 0.99$:

$P(X > 1651) = 0.990008...$ $P(X > 1652) = 0.9898$ ✓

So, $h = 1651$ ✓

(d) $H_0: \mu = 2000$, $H_1: \mu > 2000$ ✓

$\bar{X} \sim N\left(2000, \left(\frac{120}{\sqrt{30}}\right)^2\right)$ or $N(2000, 21.91^2)$ ✓

$P(\bar{X} > 2040) = 0.0340$ ✓

$0.0340 > 0.01$, so do not reject H_0. ✓

There is insufficient evidence to support the claim that the

mean lifetime of the light bulbs is more than 2000 hours. ✓

(Total for Question 3 is 10 marks)

4   The Venn diagram shows the probabilities that students in Year 12 at a particular sixth form college study chemistry (C), maths (M) and history (H).

Revision Guide
pages 127–28, 139

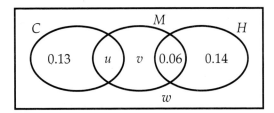

u, *v* and *w* are probabilities.

(a) If this Venn diagram represents the choices of 125 students, estimate how many take only chemistry or only history.

(2)

Watch out!

The numbers on this Venn diagram can represent frequencies, elements or probabilities. If they are numbers less than 1, they probably represent probabilities.

Given that the events *M* and *H* are independent, and that $P(M \mid C) = \frac{9}{22}$

Hint Q4b

(b) calculate the values of *u*, *v* and *w*.

(6)

Use the definitions of independent events and conditional probability to set up equations in *u* and *v*.

(a) 0.27 × 125 ✓ = 33.75 so 34 students ✓

(b) If *M* and *H* are independent, $P(M) \times P(H) = P(M \cap H)$

$(u + v + 0.06) \times 0.2 = 0.06$ ✓

$u + v + 0.06 = 0.3$

$u + v = 0.24$ ✓

$P(M \mid C) = \dfrac{P(M \cap C)}{P(C)}$

$\dfrac{9}{22} = \dfrac{u}{0.13 + u}$ ✓

$1.17 + 9u = 22u$

$1.17 = 13u$

$0.09 = u$ ✓

So: $v = 0.24 - 0.09 = 0.15$ ✓

and $w = 1 - (0.13 + 0.09 + 0.15 + 0.06 + 0.14)$

$= 1 - 0.57$

$= 0.43$ ✓

(Total for Question 4 is 8 marks)

Revision Guide
pages 131, 144

Problem solving

In part (a), you will need to use the binomial distribution twice: once for the seeds in a tray and once for the number of trays.

When using the normal distribution as an approximation for the binomial distribution, you will need to use a continuity correction. Read the question and choose the continuity correction carefully.

Modelling

Interpret your answer to part (c) in the context of the question. These answers can be quite subjective, so explain your conclusion as clearly and carefully as you can.

5   A seed company claims that 58% of its cucumber seeds germinate. A random selection of cucumber seeds is planted in 5 trays with 24 seeds in each tray.

(a) Find the probability that in at least three of the trays, 16 or more seeds will germinate.

(4)

(b) State two conditions where the normal distribution can be used as an approximation to the binomial distribution.

(1)

A random sample of 120 cucumber seeds was planted and 75 of the seeds germinated.

(c) Assuming a success rate of 58% germination, use a normal approximation to find the probability that at least 75 seeds will germinate.

(4)

(d) Does your answer to part (c) support the seed company's claim that 58% of its cucumber seeds germinate?

(2)

(a) For the seeds, $S \sim B(24, 0.58)$ ✔

   $P(S \geqslant 16) = 0.2593$ ✔

   For the trays, $T \sim B(5, 0.2593)$ ✔

   $P(T \geqslant 3) = 0.1136$ ✔

(b) n is large and p is close to 0.5 ✔

(c) $n = 120$, $p = 0.58$

   $\mu = np = 120 \times 0.58 = 69.6$ ✔

   $\sigma^2 = np(1 - p) = 120 \times 0.58 \times 0.42 = 29.232$ ✔

   $X \sim N(69.6, 29.232)$

   $P(S \geqslant 75) \approx P(X \geqslant 74.5)$ ✔ $= 0.1824$ ✔

(d) 75 is 62.5% of 120, which is a higher success rate than is

   claimed ✔, and the probability of it happening is >18%, which

   is probably sufficient to support the company's claim. ✔

(Total for Question 5 is 11 marks)

36

SECTION B: MECHANICS

Answer all questions. Write your answers in the spaces provided.

Unless otherwise indicated, whenever a numerical value of g is required, take $g = 9.8\,\mathrm{m\,s^{-2}}$ and give your answer to either 2 significant figures or 3 significant figures.

Revision Guide
page 160

LEARN IT!

Integrate the expression for acceleration with respect to time to find an expression for velocity

6 A particle is moving in a straight line. At time t seconds, $t \geqslant 0$, the acceleration of the particle, $a\,\mathrm{m\,s^{-2}}$, is given by

$$a = 2\cos\left(\frac{\pi}{6}t\right)$$

(a) The velocity of the particle at $t = 0$ is $\dfrac{2}{\pi}\mathrm{m\,s^{-1}}$.

Find an expression for the velocity of the particle at t seconds.

(3)

(b) Find the velocity of the particle when $t = 5$.

(2)

Hint Q6a

Remember to include a constant of integration.

(a) $a = 2\cos\left(\dfrac{\pi}{6}t\right)$

$v = \displaystyle\int a\,dt = \int 2\cos\frac{\pi}{6}t\,dt$ ✔

$= \dfrac{12}{\pi}\sin\left(\dfrac{\pi}{6}t\right) + c$ ✔

When $t = 0$, $v = \dfrac{2}{\pi}$, so $\dfrac{2}{\pi} = 0 + c$, so $c = \dfrac{2}{\pi}$

So, $v = \dfrac{12}{\pi}\sin\left(\dfrac{\pi}{6}t\right) + \dfrac{2}{\pi}$ ✔

(b) When $t = 5$, $v = \dfrac{12}{\pi}\sin\left(\dfrac{5\pi}{6}\right) + \dfrac{2}{\pi}$

$= \dfrac{12}{\pi} \times \dfrac{1}{2} + \dfrac{2}{\pi}$ ✔

$= \dfrac{8}{\pi}\mathrm{m\,s^{-1}}$ ✔

(Total for Question 6 is 5 marks)

37

Revision Guide
pages 160, 155

Hint Q7c

After differentiating to find the acceleration vector, you need to use **F** = m**a**, which will give **F** as a vector. The question asks for the magnitude of **F**, so there is still work to do.

7   A particle, P, of mass 0.6 kg moves under the action of a single force **F** newtons.

At time t seconds, the velocity, **v** m s^{-1}, of P is given by

$$\mathbf{v} = 4t^2\mathbf{i} + (5t - 1)\mathbf{j}$$

Find:

(a) the time after which the particle is moving parallel to the vector **i**

(1)

(b) the acceleration of P at time t seconds

(2)

(c) the magnitude of **F** when $t = 1.5$.

(3)

(a) $5t - 1 = 0$ so $t = 0.2$ ✓

(b) $\underset{\sim}{\mathbf{v}} = 4t^2\underset{\sim}{\mathbf{i}} + (5t - 1)\underset{\sim}{\mathbf{j}}$

$\underset{\sim}{\mathbf{a}} = \dfrac{d\underset{\sim}{\mathbf{v}}}{dt} = 8t\underset{\sim}{\mathbf{i}} + 5\underset{\sim}{\mathbf{j}}$ ✓✓

(c) When $t = 1.5$, $\underset{\sim}{\mathbf{a}} = 12\underset{\sim}{\mathbf{i}} + 5\underset{\sim}{\mathbf{j}}$

Using $\underset{\sim}{\mathbf{F}} = m\underset{\sim}{\mathbf{a}}$:

$\underset{\sim}{\mathbf{F}} = 0.6(12\underset{\sim}{\mathbf{i}} + 5\underset{\sim}{\mathbf{j}})$

$= 7.2\underset{\sim}{\mathbf{i}} + 3\underset{\sim}{\mathbf{j}}$ ✓

So $|\underset{\sim}{\mathbf{F}}| = \sqrt{7.2^2 + 3^2}$ ✓

$= 7.8$N ✓

(Total for Question 7 is 6 marks)

38

8 The diagram shows two
   particles, *A* of mass 6 kg and
   B of mass 2.5 kg, connected by
   a light inextensible string
   passing over a smooth pulley.

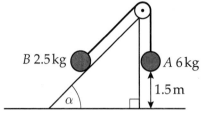

B 2.5 kg   A 6 kg
             1.5 m

Initially *B* is held at rest on a
rough inclined plane, inclined
at an angle α to the horizontal, where $\tan \alpha = \frac{4}{3}$.

The coefficient of friction between *B* and the inclined plane is $\frac{1}{6}$.

The system is released from rest with *A* at a height 1.5 m above the
ground.

(a) For the motion until *A* hits the ground, work out the
     acceleration of the system.

     (7)

(b) *A* does not rebound when it hits the ground and *B* continues
     moving up the inclined plane. Given that *B* does not reach
     the pulley, find how much further *B* travels before coming to
     instantaneous rest.

     (5)

(c) State how you have used the fact that

     (i) the string is inextensible

     (ii) the pulley is smooth

     in your calculations.

     (2)

Problem solving

In parts (a) and (b) you
need to think carefully
about the forces
acting on the particles.
Use components
of the 2.5 kg mass,
both parallel and
perpendicular to the
inclined plane. The
movement is up the
plane in both parts, so
friction acts down the
inclined plane.

Draw clear diagrams
and show all the forces
acting on the particles.

The question involves
resolving forces, using
$F = ma$ and the *suvat*
formulae.

Hint Q8a

Use $F = ma$ for both
particles and solve
simultaneously to find *a*.

Hint Q8b

Once *A* has hit the
ground, the only forces
acting on *B* are its
weight components and
friction.

Use $F = ma$ and one of
the *suvat* formulae.

(a)

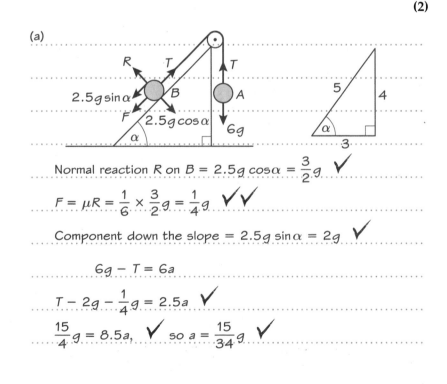

Normal reaction R on $B = 2.5g \cos\alpha = \frac{3}{2}g$ ✓

$F = \mu R = \frac{1}{6} \times \frac{3}{2}g = \frac{1}{4}g$ ✓✓

Component down the slope $= 2.5g \sin\alpha = 2g$ ✓

$6g - T = 6a$

$T - 2g - \frac{1}{4}g = 2.5a$ ✓

$\frac{15}{4}g = 8.5a,$ ✓ so $a = \frac{15}{34}g$ ✓

(b) Using $v^2 = u^2 + 2as$:

$v^2 = 0 + 2 \times \dfrac{15}{34}g \times 1.5$, so $v^2 = \dfrac{45}{34}g$ ✓

Forces on B, parallel to slope $= 2g + \dfrac{1}{4}g = \dfrac{9}{4}g$ downwards.

Using $F = ma$ on B:

$-\dfrac{9}{4}g = 2.5a$ ✓, so $a = -\dfrac{9}{10}g$ ✓

Using $v^2 = u^2 + 2as$ on B, where $u^2 = \dfrac{45}{34}g$:

$0 = \dfrac{45}{34}g - 2 \times \dfrac{9}{10}g \times s$ ✓

$s = \dfrac{225}{306} = 0.735\,\text{m}$ (3 s.f) ✓

(c) (i) Both particles move with the same acceleration. ✓

(ii) Tension is the same on both sides of the pulley. ✓

(Total for Question 8 is 14 marks)

40

9   A basketball player is taking two free-throw shots. He stands 4.2 m from the centre of the hoop of the basket. The basket is 3 m above the ground and the player releases the basketball from a height of 2 m.

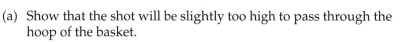

For his first shot, the basketball is projected at an angle of 50° with a speed of 7.5 m s⁻¹.

(a) Show that the shot will be slightly too high to pass through the hoop of the basket.

(6)

For his second shot, the player keeps the angle of projection at 50° but adjusts the speed of the throw. This time, the basketball passes through the centre of the hoop and he scores one point.

(b) Find the new speed of projection.

(6)

(a) Horizontally: time = $\dfrac{\text{distance}}{\text{speed}}$

$= \dfrac{4.2}{7.5 \cos 50°} = 0.8712...\text{s}$ ✔✔

Vertically, using $s = ut + \dfrac{1}{2}at^2$:

$h = (7.5 \sin 50°)(0.8712...) - \dfrac{1}{2} \times 9.8 \times (0.8712...)^2$ ✔✔

where h is the height above 2 m.

Giving $h = 1.2862...\,\text{m}$ ✔

So the basketball will hit 0.2862...m above the hoop, and

so will not pass through the hoop. ✔

(b) Horizontally: time = $\dfrac{\text{distance}}{\text{speed}} = \dfrac{4.2}{u \cos 50°}$ ✔

where u is the required speed.

Vertically, using $s = ut + \dfrac{1}{2}at^2$, where $s = 1$: ✔

$1 = u \sin 50°\left(\dfrac{4.2}{u \cos 50°}\right) - \dfrac{1}{2} \times 9.8 \times \left(\dfrac{4.2}{u \cos 50°}\right)^2$ ✔✔

$= 4.2 \tan 50° - \dfrac{4.9 \times 4.2^2}{u^2 \cos^2 50°}$

$\dfrac{4.9 \times 4.2^2}{u^2 \cos^2 50°} = 4.2 \tan 50° - 1$ ✔

Giving $u = 7.23\,\text{m s}^{-1}$ (3 s.f) ✔

(Total for Question 9 is 12 marks)

Revision Guide
pages 170, 171

Problem solving

In both parts, you need to establish the time for the basketball to reach the hoop, using the horizontal component of the speed. Then use this information, along with a *suvat* formula, for the vertical motion.

Hint Q9b

You are effectively using the equation of the trajectory. This is a particular case of that concept. The equation for u (the required speed) looks a little complicated, so be careful when isolating the term in u^2, which leads to the solution.

Modelling

You are modelling the basketball as a particle. Always interpret your solutions in the context of the problem.

41

Revision Guide
page 174

Problem solving

Draw a diagram for both parts of this question. Show normal reactions, frictional forces and the weights of the ladder and the man standing on it.

Hint Q10a

Resolve horizontally and vertically and choose a point to take moments about.

Remember, moment = force × perpendicular distance.

Hint Q10b

There is an extra force (kW) acting vertically downwards at A and the man is now standing at the top of the ladder. Use the value of μ from part (a), and resolve and take moments as before. If you consider friction to be limiting, you will be calculating the minimum value of k.

Modelling

The man is modelled as a particle. The ladder is uniform, so its weight will act at its midpoint.

10

A ladder, AB, of weight W and length $4a$, has one end on rough horizontal ground. The coefficient of friction is μ. The other end rests against a smooth vertical wall and the ladder makes an angle α with the ground, where $\tan \alpha = \frac{7}{2}$.

A man of weight $6W$ stands at C on the ladder, where $AC = 3a$. The ladder is modelled as a uniform rod, lying in a vertical plane, and the man is modelled as a particle. The system is in limiting equilibrium.

(a) Show that $\mu = \frac{10}{49}$

(6)

A downward force, equivalent to a weight of kW, is applied at A, enabling the man to climb to the top of the ladder at B. The system is in equilibrium.

(b) Find the range of possible values of k.

(7)

(a)

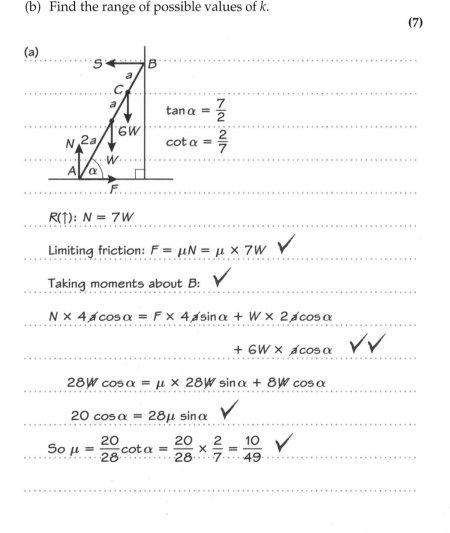

$\tan \alpha = \frac{7}{2}$

$\cot \alpha = \frac{2}{7}$

$R(\uparrow): N = 7W$

Limiting friction: $F = \mu N = \mu \times 7W$ ✓

Taking moments about B: ✓

$N \times 4a\cos\alpha = F \times 4a\sin\alpha + W \times 2a\cos\alpha$

$\qquad\qquad\qquad\qquad + 6W \times a\cos\alpha$ ✓✓

$28W\cos\alpha = \mu \times 28W\sin\alpha + 8W\cos\alpha$

$20\cos\alpha = 28\mu\sin\alpha$ ✓

So $\mu = \frac{20}{28}\cot\alpha = \frac{20}{28} \times \frac{2}{7} = \frac{10}{49}$ ✓

42

(b)

Assume limiting friction, so k will be a minimum.

R($\uparrow$): $P = (7 + k)W$

$F' = \mu P = \dfrac{10}{49}(7 + k)W$ ✓

R($\rightarrow$): $Q = F' = \dfrac{10}{49}(7 + k)W$ ✓

Taking moments about A: ✓

$\dfrac{10}{49}(7 + k)W \times 4a\sin\alpha = 6W \times 4a\cos\alpha + W \times 2a\cos\alpha$ ✓

$\dfrac{40}{49}(7 + k)\sin\alpha = 26\cos\alpha$ ✓

$7 + k = \dfrac{26 \times 49}{40}\cot\alpha$

$= \dfrac{26 \times 49}{40} \times \dfrac{2}{7} = \dfrac{91}{10} = 9.1$

so $k = 2.1$ ✓

So, for equilibrium, $k \geqslant 2.1$ ✓

(Total for Question 10 is 13 marks)

TOTAL FOR PAPER IS 100 MARKS

Revision Guide
page 87

LEARN IT!

The small angle
approximations are:

$\sin\theta \approx \theta$

$\tan\theta \approx \theta$

$\cos\theta \approx 1 - \dfrac{\theta^2}{2}$

Hint Q1b

Small angle
approximations only
work when the angle is
given in radians.

Paper 1: Pure Mathematics 1

Answer all questions. Write your answers in the spaces provided.

1   (a)   Given that θ is small, use the small angle approximations for
$\cos\theta$ and $\sin\theta$ to show that

$$\frac{1 - \cos^2 3\theta}{3\theta \sin 2\theta} \approx \frac{3}{2}$$

(3)

Beth uses $\theta = 5°$ to test the approximation in part (a). This is her
working:

> Using my calculator:
> $$\frac{1 - \cos^2 15°}{15 \times \sin 10°} = 0.0257...$$
> So the approximation is not true for $\theta = 5°$

(b)   Identify the mistake in Beth's working, and show that the
approximation is accurate to 1 decimal place.

(2)

(a) Numerator $= 1 - \left(1 - \dfrac{(3\theta)^2}{2}\right)^2$

$= 1 - \left(1 - \dfrac{9\theta^2}{2}\right)^2$

$= 1 - \left(1 - 9\theta^2 + \dfrac{81\theta^4}{4}\right)$

$\approx 9\theta^2$ ✔

Denominator $= 3\theta \times 2\theta = 6\theta^2$ ✔

So: $\dfrac{\text{Numerator}}{\text{Denominator}} = \dfrac{9\theta^2}{6\theta^2} = \dfrac{3}{2}$ ✔

(b) Beth has used degrees, instead of radians.

$5° = \dfrac{5\pi}{180} = \dfrac{\pi}{36}$ rad, so use $\theta = \dfrac{\pi}{36}$ ✔

$1 - \cos^2 3\theta = 1 - \cos^2\left(\dfrac{\pi}{12}\right) = 0.06698...$

$3\theta \sin 2\theta = \dfrac{\pi}{12} \times \sin\dfrac{\pi}{18} = 0.0454...$ } Dividing $= 1.4735...$

≈ 1.5 (1 d.p.) ✔

(Total for Question 1 is 5 marks)

44

2 The function f is defined by f: $x \to |2x - 7|$, $x \in \mathbb{R}$.

(a) Sketch the graph with equation $y = f(x)$, showing the coordinates of the points where the graph cuts or meets the axes.

(2)

(b) Solve $f(x) = 20 + x$

(3)

Revision Guide
page 66

Hint Q2b

There will be two solutions. Consider the positive and negative arguments.

Problem solving

You could use your sketch to check that your solutions look about right.

(a)

✔ for shape

✔ for points (0, 7) and (3.5, 0)

(b) $2x - 7 = 20 + x$

$x = 27$ ✔

$-(2x - 7) = 20 + x$ ✔

$-2x + 7 = 20 + x$

$x = \dfrac{-13}{3}$ ✔

(Total for Question 2 is 5 marks)

45

136

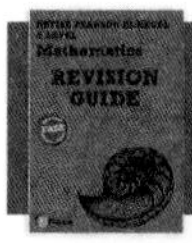

Problem solving

This is called the ambiguous case because there are two possible triangles that work with these values. Sketch the triangle then use the sine rule.

LEARN IT!

$$\frac{\sin A}{a} = \frac{\sin B}{b}$$

3 In the triangle ABC, $AB = 8\,cm$, $AC = 11\,cm$ and angle $ACB = \frac{\pi}{9}$ radians.

Find the two possible values of angle ABC.

(5)

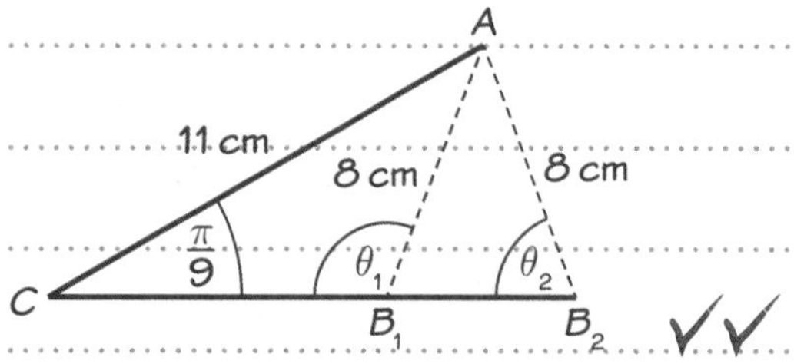

$$\frac{\sin \theta}{11} = \frac{\sin \frac{\pi}{9}}{8} \checkmark$$

$\theta = 0.49$ radians $\checkmark$ or 2.65 radians $\checkmark$

(Total for Question 3 is 5 marks)

46

4   $f(x) = e^{x\sqrt{3}} \cos x$, $\dfrac{-\pi}{2} \leqslant x \leqslant \dfrac{\pi}{2}$

The diagram shows a sketch of the curve C with equation $y = f(x)$

Revision Guide
pages 39/40, 90, 92

(a) Find the x-coordinate of the stationary point P, on C.

Give your answer as a multiple of π.

(4)

Hint Q4a

Use the product rule, then solve a trig equation for the x-coordinate of P.

(b) Find the equation of the tangent to C at the point where $x = 0$.

(3)

Hint Q4b

You will need to find the values of y and $\dfrac{dy}{dx}$ when $x = 0$.

(a) $y = e^{x\sqrt{3}} \cos x$

$\dfrac{dy}{dx} = e^{x\sqrt{3}}(-\sin x) + \sqrt{3}\, e^{x\sqrt{3}} \cos x$ ✔✔

$= e^{x\sqrt{3}}(\sqrt{3} \cos x - \sin x)$

$\dfrac{dy}{dx} = 0$ when $\sqrt{3} \cos x = \sin x$ ✔

$\tan x = \sqrt{3}$

$x = \dfrac{\pi}{3}$

So the x-coordinate of P is $\dfrac{\pi}{3}$ ✔

(b) When $x = 0$, $\dfrac{dy}{dx} = e^{0}(\sqrt{3} \cos 0 - \sin 0) = \sqrt{3}$ ✔

and $y = e^{0} \cos 0 = 1$ ✔

Equation of the tangent is

$y - 1 = \sqrt{3}\, x$ or $y = \sqrt{3}\, x + 1$ ✔

(Total for Question 4 is 7 marks)

47

138

Revision Guide
page 56

5   Use proof by contradiction to show that there are no rational numbers which satisfy the equation

$$x^3 - 3x + 1 = 0$$

(5)

Problem solving

Assume that there exists a rational solution. You can write this in the form $x = \dfrac{p}{q}$, where p and q are integers with no common factors. Substitute for x, and multiply through by q^3 to clear all the denominators.

Problem solving

Consider all combinations of p and q being odd or even. There are four possible cases to consider.

Assume that $x = \dfrac{p}{q}$, in its simplest form, where p and q are both integers, is a solution. ✓

Then $\dfrac{p^3}{q^3} - 3\dfrac{p}{q} + 1 = 0$

Multiplying through by q^3 gives

$p^3 - 3pq^2 + q^3 = 0$ ✓

If p and q are both odd:

p^3 is odd, $3pq^2$ is odd and q^3 is odd.

So LHS is odd, but RHS = 0 is not odd, so there is

a contradiction.

If p is even and q is odd:

p^3 is even, $3pq^2$ is even and q^3 is odd. So LHS is odd,

but RHS = 0 is not odd, so there is

a contradiction.   ✓✓

If p is odd and q is even:

p^3 is odd, $3pq^2$ is even and q^3 is even.

So LHS is odd, but RHS = 0 is not odd, so there is

a contradiction.

p and q cannot both be even because $\dfrac{p}{q}$ was assumed to be in its

simplest form. ✓

So there are no rational number solutions.

(Total for Question 5 is 5 marks)

48

6 A circle intersects the *x*-axis at the points *A*(2, 0) and *B*(8, 0).
The *y*-axis is a tangent to the circle at *C*.

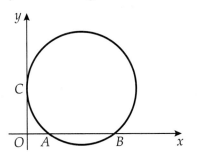

(a) Work out the equation of the circle.

(4)

(b) The point *E* is on the circle vertically above *B*. Find the equation
of the tangent at the point *E* in the form $ax + by = c$.

(5)

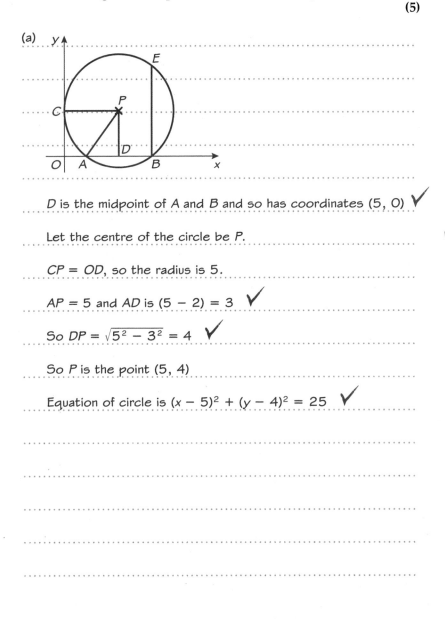

(a)

D is the midpoint of *A* and *B* and so has coordinates (5, 0) ✓

Let the centre of the circle be *P*.

CP = *OD*, so the radius is 5.

AP = 5 and *AD* is (5 − 2) = 3 ✓

So $DP = \sqrt{5^2 - 3^2} = 4$ ✓

So *P* is the point (5, 4)

Equation of circle is $(x - 5)^2 + (y - 4)^2 = 25$ ✓

Revision Guide
pages 20, 21

Hint Q6a

The midpoint of *AB* is the *x*-coordinate of the centre.

Use Pythagoras to find the *y*-coordinate of the centre of the circle.

Hint Q6b

Find the coordinates of *E*. Find the gradient of the line that connects *E* and the centre of the circle. Use this to find the gradient of the tangent and then use $y - y_1 = m(x - x_1)$ with the coordinates of *E* to find the equation of the tangent.

LEARN IT!

The gradient of the tangent is the negative reciprocal of the radius.

(b) Triangle ADP is similar to triangle ABE.

E has coordinates $(8, 8)$ ✓

Gradient $EP = \dfrac{8-4}{8-5} = \dfrac{4}{3}$ ✓

Gradient of tangent at $E = \dfrac{-3}{4}$ ✓

Equation of tangent is

$y - y_1 = m(x - x_1)$

$y - 8 = \dfrac{-3}{4}(x - 8)$ ✓

$4y - 32 = -3x + 24$

$3x + 4y = 56$ ✓

(Total for Question 6 is 9 marks)

50

141

7 Solve, for $-\pi < x < \pi$, the equation

$$6\sin^2 x + \cos x - 4 = 0$$

giving your answers to 2 decimal places.

(5)

$6\sin^2 x + \cos x - 4 = 0$

$6(1 - \cos^2 x) + \cos x - 4 = 0$ ✓

$6\cos^2 x - \cos x - 2 = 0$ ✓

$(2\cos x + 1)(3\cos x - 2) = 0$ ✓

$\cos x = \dfrac{-1}{2}$ or $\dfrac{2}{3}$ ✓

So $x = 2.09$ rad, -2.09 rad, 0.84 rad, -0.84 rad ✓

Revision Guide
page 32

Hint

$\sin^2 x = 1 - \cos^2 x$.
Simplify the resulting
quadratic and factorise.

Watch out!

There are four solutions
in total.

(Total for Question 7 is 5 marks)

51

142

Hint Q8a

Write the expression as $(x + 1)(2 - 3x)^{-2}$ then take a common factor of 2 out of the second bracket.

When 2 is taken out as a common factor it becomes 2^{-2}. Expand the second bracket to the x^2 term, then multiply by the first bracket.

LEARN IT!

Remember:

$(a + bx)^n = a^n\left(1 + \dfrac{bx}{a}\right)^n$

and the expression is valid for

$\left|\dfrac{bx}{a}\right| < 1$ or

$|x| < \dfrac{a}{b}$

8 (a) Expand

$$\frac{x + 1}{(2 - 3x)^2}$$

in ascending powers of x, up to and including the term in x^2, giving each term as a simplified fraction.

(5)

(b) State the range of values of x for which the expansion converges.

(1)

(a) $\dfrac{x + 1}{(2 - 3x)^2} = (x + 1)(2 - 3x)^{-2}$ ✓

$= (x + 1)\left[(2)^{-2}\left(1 - \dfrac{3}{2}x\right)^{-2}\right]$ ✓

$= \dfrac{1}{4}(x + 1)\left[1 + (-2)\left(-\dfrac{3}{2}x\right) + \dfrac{(-2)(-3)}{2!}\left(\dfrac{-3}{2}x\right)^2\right]$ ✓

$= \dfrac{1}{4}(x + 1)\left(1 + 3x + \dfrac{27}{4}x^2\right)$ ✓

$= \dfrac{1}{4}\left(x + 3x^2 + \dfrac{27}{4}x^3 + 1 + 3x + \dfrac{27}{4}x^2\right)$

$\approx \dfrac{1}{4}\left(1 + 4x + \dfrac{39}{4}x^2\right)$ ✓

(b) $|x| < \dfrac{2}{3}$ ✓

(Total for Question 8 is 6 marks)

52

143

9 A liquid is being heated. At time t seconds, the temperature of the liquid is θ °C.

Revision Guide
pages 41, 97, 109

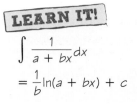

The rate of increase of the temperature of the liquid is modelled by the differential equation

$$\frac{d\theta}{dt} = k(170 - 2\theta)$$

where k is a positive constant.

Hint Q9a

Separate the variables then integrate. Don't forget the constant of integration, which must be evaluated.

(a) Given that $\theta = 20$ when $t = 0$, show that $\theta = A - B\,e^{-2kt}$ where A and B are integers to be found.

(6)

(b) Given that $k = 0.004$, find the time for the temperature of the liquid to reach 70 °C.

(2)

LEARN IT!

$$\int \frac{1}{a + bx}\,dx$$
$$= \frac{1}{b}\ln(a + bx) + c$$

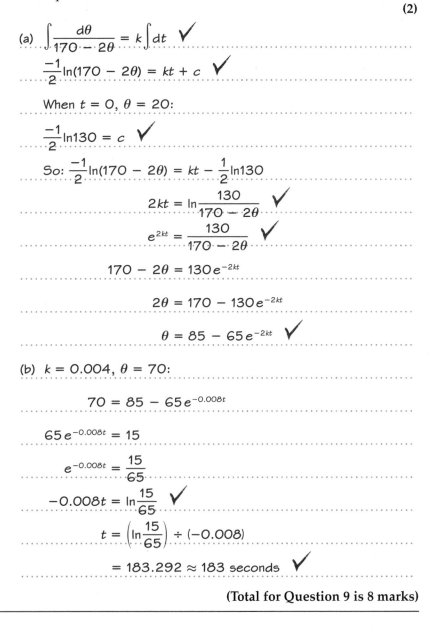

(a) $\displaystyle\int \frac{d\theta}{170 - 2\theta} = k\int dt$ ✓

$\dfrac{-1}{2}\ln(170 - 2\theta) = kt + c$ ✓

When $t = 0$, $\theta = 20$:

$\dfrac{-1}{2}\ln 130 = c$ ✓

So: $\dfrac{-1}{2}\ln(170 - 2\theta) = kt - \dfrac{1}{2}\ln 130$

$2kt = \ln\dfrac{130}{170 - 2\theta}$ ✓

$e^{2kt} = \dfrac{130}{170 - 2\theta}$ ✓

$170 - 2\theta = 130\,e^{-2kt}$

$2\theta = 170 - 130\,e^{-2kt}$

$\theta = 85 - 65\,e^{-2kt}$ ✓

(b) $k = 0.004$, $\theta = 70$:

$70 = 85 - 65\,e^{-0.008t}$

$65\,e^{-0.008t} = 15$

$e^{-0.008t} = \dfrac{15}{65}$

$-0.008t = \ln\dfrac{15}{65}$ ✓

$t = \left(\ln\dfrac{15}{65}\right) \div (-0.008)$

$= 183.292 \approx 183$ seconds ✓

(Total for Question 9 is 8 marks)

Revision Guide
pages 90, 98

Hint Q10a

Use the product rule and equate to zero.

Hint Q10a

Write out the product rule and your expressions for u and v before substituting.

LEARN IT!

You can learn the product rule for a function uv as $u'v + v'u$.

Hint Q10d

Make sure f(x) changes sign either side of P.

10

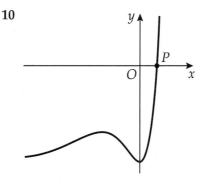

The diagram shows a sketch of the graph of

$$f(x) = 9x^2e^{2x} - 4, x \in \mathbb{R}$$

(a) Find the exact coordinates of the turning points on this curve.

(5)

(b) Show that the equation f(x) = 0 can be written in the form

$$x = \pm\frac{2}{3}e^{-x}$$

(1)

The equation f(x) has a root α, where $\alpha = 0.4$, to 1 decimal place.

(c) Starting with $x_0 = 0.4$, use the iterative formula

$$x_{n+1} = \frac{2}{3}e^{-x_n}$$

to calculate the values of x_1, x_2 and x_3, giving your answers to 4 decimal places.

(3)

(d) Give an accurate estimate to 2 decimal places of the x-coordinate of P, justifying your answer.

(2)

(a) $y = f(x) = 9x^2e^{2x} - 4$

$\dfrac{dy}{dx} = 9(2e^{2x}x^2 + 2xe^{2x})$

$= 18xe^{2x}(x + 1)$ ✓✓

Turning points at $x = 0$ and $x = -1$ ✓

When $x = 0$: $y = -4$ ✓

When $x = -1$: $y = 9(1)e^{-2} - 4 = \dfrac{9}{e^2} - 4$ ✓

So the coordinates of the turning points are $(0, -4)$

and $\left(-1, \dfrac{9}{e^2} - 4\right)$

54

(b) $9x^2 e^{2x} = 4$

$$x^2 = \frac{4}{9} e^{-2x}$$

$$x = \pm \frac{2}{3} e^{-x} \quad \checkmark$$

(c) $x_1 = 0.446880\ldots \approx 0.4469 \quad \checkmark$

$x_2 = 0.426413\ldots \approx 0.4264 \quad \checkmark$

$x_3 = 0.435230\ldots \approx 0.4352 \quad \checkmark$

(d) Testing for a sign change:

$\left.\begin{array}{l} f(0.425) = -0.1966\ldots \\[4pt] f(0.435) = 0.0649\ldots \end{array}\right\} \quad \checkmark$

x-coordinate of P is 0.43 (2 d.p.) $\quad \checkmark$

(Total for Question 10 is 11 marks)

Hint Q11a

The information given in the question tells you the coordinates of the turning point and one other point on the parabola.

Modelling

You can give any reasonable answer but make sure you refer to the context of the question.

11 The diagram shows the trajectory of an arrow that is fired horizontally from the top of a castle.

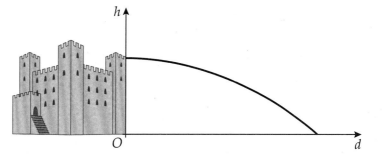

The point of projection is 20 m above ground level, and the arrow hits the ground a horizontal distance of 100 m from the base of the castle.

(a) Given that the trajectory of the arrow can be modelled as a parabola, find an equation for the height of the arrow above the ground, h, in terms of the horizontal distance travelled, d.

Give your answer in the form $h = ad^2 + bd + c$, and specify a suitable range of values for d.

(4)

(b) State one limitation of the model.

(1)

(a) Turning point is at (0, 20) so equation has form

$h = ad^2 + 20$ ✓

When $d = 100$, $h = 0$, so $0 = 10000a + 20$ ✓

so $a = -0.002$ ✓

So $h = -0.002d^2 + 20$, $0 \leqslant d \leqslant 100$ ✓

(b) The path of the arrow is unlikely to be a perfect parabola. ✓

(Total for Question 11 is 5 marks)

12 $\sum_{r=1}^{k} (3 + \frac{1}{2}r) = 105$

Find the value of k.

(5)

Sequence terms are 3.5, 4, 4.5, 5, ...

First term $= a = 3.5$, common difference $= d = 0.5$

$S_n = \frac{1}{2}n(2a + (n-1)d)$ ✓

$S_k = \frac{1}{2}k\left(7 + \frac{1}{2}(k-1)\right)$ ✓

$\frac{1}{2}k\left(7 + \frac{1}{2}(k-1)\right) = 105$

$14k + k^2 - k = 420$ ✓

$k^2 + 13k - 420 = 0$

$(k + 28)(k - 15) = 0$

$k = -28$ or 15 ✓

So $k = 15$ as k must be a positive number. ✓

Revision Guide
pages 68, 72

Problem solving

Write out the first few terms of the arithmetic series with general term $3 + \frac{1}{2}r$, then write an expression for the sum to k terms.

Hint

k must be a positive integer.

(Total for Question 12 is 5 marks)

Revision Guide
pages 90, 103, 108

Hint Q13a

Two of the answers are given so you can check that you are doing the correct operation on your calculator. Check that your values look to fit the shape of the curve in the diagram.

Hint Q13b

The trapezium rule is

$$\frac{1}{2}h[(y_0 + y_n) + 2(y_1 + y_2 + \ldots + y_{n-1})]$$

where $h = \dfrac{b - a}{n}$.

This is given in the formulae book that you are provided with in your examination.

Hint Q13c

Differentiate $\cot x$ to get dx in terms of x and du. Substitute $\cot x$ and dx for terms with u and cancel out the cosec^2 terms.

Find u when $x = \dfrac{\pi}{6}$ and $x = \dfrac{\pi}{3}$, and substitute in for the limits. Then integrate with respect to u and find the exact value by manipulating the surds.

13

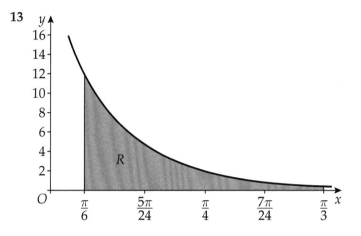

The diagram shows the graph of the curve with equation

$$y = \operatorname{cosec}^2 x \cot^2 x, \quad \frac{\pi}{6} \leqslant x \leqslant \frac{\pi}{3}$$

The finite region R is bounded by the lines $x = \dfrac{\pi}{6}$, $x = \dfrac{\pi}{3}$, the x-axis and the curve as shown.

(a) Complete the table with the values of y corresponding to $x = \dfrac{5\pi}{24}, \dfrac{\pi}{4}$ and $\dfrac{7\pi}{24}$

x	$\dfrac{\pi}{6}$	$\dfrac{5\pi}{24}$	$\dfrac{\pi}{4}$	$\dfrac{7\pi}{24}$	$\dfrac{\pi}{3}$
y	12				0.44444

(2)

(b) Use the trapezium rule with the values in the table to find an approximate value for the area of R, giving your answer to 3 significant figures.

(4)

(c) By using the substitution $u = \cot x$ and integrating, find the exact value of the area of R.

(7)

(a)

x	$\dfrac{\pi}{6}$	$\dfrac{5\pi}{24}$	$\dfrac{\pi}{4}$	$\dfrac{7\pi}{24}$	$\dfrac{\pi}{3}$
y	12	4.58295	2	0.93547	0.44444

✓✓   (any two correct gains 1 mark, all three gains 2 marks)

(b)   Area $\approx \dfrac{1}{2} \times \dfrac{\pi}{24}$ [(12 + 0.44444) +

2(4.58295 + 1 + 0.93547)]   ✓✓

= 1.79864557...   ✓

= 1.80 (3 s.f.)   ✓

(c) $u = \cot x$

$\dfrac{du}{dx} = -\operatorname{cosec}^2 x$ ✓

$dx = \dfrac{du}{-\operatorname{cosec}^2 x}$

$\operatorname{cosec}^2 x \cot^2 x \, dx = \operatorname{cosec}^2 x \,(u^2)\,\dfrac{du}{-\operatorname{cosec}^2 x} = -u^2 \, du$ ✓

When $x = \dfrac{\pi}{6}$, $u = \sqrt{3}$

When $x = \dfrac{\pi}{3}$, $u = \dfrac{1}{\sqrt{3}}$ ✓

Area $= \displaystyle\int_{\frac{\pi}{6}}^{\frac{\pi}{3}} \operatorname{cosec}^2 x \cot^2 x \, dx$

$= \displaystyle\int_{\sqrt{3}}^{\frac{1}{\sqrt{3}}} -u^2 \, du$ ✓

$= \left[\dfrac{-u^3}{3} \right]_{\sqrt{3}}^{\frac{1}{\sqrt{3}}}$ ✓

$= -\dfrac{\left(\dfrac{1}{\sqrt{3}}\right)^3}{3} - \left(-\dfrac{\sqrt{3}^3}{3} \right)$ ✓

$= \dfrac{-1}{9\sqrt{3}} + \dfrac{3\sqrt{3}}{3}$

$= \dfrac{-\sqrt{3}}{27} + \sqrt{3}$

$= \dfrac{26\sqrt{3}}{27}$ ✓

(Total for Question 13 is 13 marks)

Hint Q14b, c

Use the result from part (a) and consider the significance of $\cos(\theta + \alpha)$ when the cabin is at its highest.

Hint Q14d

Find the two values of t and work out the length of time between them.

14 (a) Express $35\cos\theta - 12\sin\theta$ in the form $R\cos(\theta + \alpha)$, where $R > 0$ and $0 < \alpha < \frac{\pi}{2}$.
State the value of R and give the value of α to 4 decimal places.

(3)

An amusement park has a big-wheel ride at the entrance. The height above the ground of one of the cabins is modelled by the equation

$$h = 39 - 35\cos0.25t + 12\sin0.25t, \ t > 0$$

where h is the height of the cabin, in metres, and t is the number of minutes after a visitor boards the cabin. The angles are measured in radians.

Find:

(b) the maximum height reached by the cabin

(2)

(c) the time taken for a cabin to reach its maximum height.

(2)

The best panoramic views are obtained when the cabins are 50 m or more above the ground.

(d) Calculate the number of minutes that a cabin is at least 50 m above the ground in each revolution of the big wheel.

(4)

(a) $35\cos\theta - 12\sin\theta \equiv R\cos\theta\cos\alpha - R\sin\theta\sin\alpha$

$\left.\begin{array}{l} R\sin\alpha = 12 \\ R\cos\alpha = 35 \end{array}\right\}$ ✓

$\tan\alpha = \dfrac{12}{35}$

$\alpha = 0.3303...$ ✓

$R = \sqrt{12^2 + 35^2} = 37$ ✓

So $35\cos\theta - 12\sin\theta = 37\cos(\theta + 0.3303...)$

(b) $h = 39 - 35\cos0.25t + 12\sin0.25t$

$= 39 - (35\cos0.25t - 12\sin0.25t)$

$= 39 - 37\cos(0.25t + 0.3303...)$

Maximum height occurs when $\cos(0.25t + 0.3303...) = -1$ ✓

So $h_{max} = 39 + 37 = 76$m ✓

(c) $\cos(0.25t + 0.3303...) = -1$ ✓

$0.25t + 0.3303... = \pi$ ✓

So time to maximum height = 11.2 minutes (3 s.f.) ✓

(d) Solve $39 - 37\cos(0.25t + 0.3303...) = 50$:

$\cos(0.25t + 0.3303...) = \dfrac{-11}{37}$ ✓

$0.25t + 0.3303... = 1.87265...$ or $4.41052...$ ✓

giving $t = 6.17$ minutes and 16.32 minutes ✓

So time above 50m = $16.32 - 6.17$

$= 10.15$ minutes = 10 minutes 9 seconds ✓

(Total for Question 14 is 11 marks)

TOTAL FOR PAPER IS 100 MARKS

Revision Guide
page 60

Paper 2: Pure Mathematics 2

Answer all questions. Write your answers in the spaces provided.

1   $f(x) = e^{2x} + 2, \quad x \in \mathbb{R}$

   $g(x) = \ln(x - 2), \quad x > 2, x \in \mathbb{R}$

   (a) Write down the composite functions

       (i) $fg(x)$

       (ii) $gf(x)$

       simplifying your answers.

   (3)

   (b) Hence find the solution to $fg(x) = gf(x)$, and justify that it is unique.

   (2)

Hint Q1a

Make sure you get the order of the functions correct, and don't forget to simplify your answers.

Hint Q1b

A value of x can only be the solution to $fg(x) = gf(x)$ if it lies within the domain of both functions.

(a) (i) $fg(x) = e^{2\ln(x-2)} + 2$ ✓

         $= e^{\ln(x-2)^2} + 2$

         $= (x - 2)^2 + 2$ or $x^2 - 4x + 6$ ✓

    (ii) $gf(x) = \ln(e^{2x} + 2 - 2)$

         $= \ln e^{2x} = 2x$ ✓

(b) $x^2 - 4x + 6 = 2x$

    $x^2 - 6x + 6 = 0$ ✓

    $x = \dfrac{6 \pm \sqrt{12}}{2} = 3 \pm \sqrt{3}$

    So $x = 3 + \sqrt{3}$ since $3 - \sqrt{3}$ is outside the domain of g ✓

(Total for Question 1 is 5 marks)

2

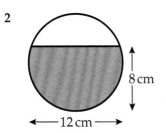

The diagram shows the circular cross-section of a water pipe with diameter 12 cm. Water flows through the pipe at a depth of 8 cm.

(a) Find the area of the water in the cross-section of the pipe to 3 significant figures.

(6)

(b) The water flows through the pipe at $0.4\,\text{ms}^{-1}$. How many litres of water flow through the pipe each hour?

(2)

Revision Guide pages 76, 77

Problem solving

Drawing your own sketch (or adding to the sketch given in the question) will help you here.

LEARN IT!

Area of a sector
$= \frac{1}{2}r^2\,\theta$, (θ in radians)

Area of a triangle
$= \frac{1}{2}\,ab\sin C$

Hint Q2b

Be careful with the units.

(a)

Radius = 6 cm so height of triangle = 2 cm ✓

$\cos\theta = \dfrac{2}{6}$

$\theta = 1.23095...$ rad ✓

Area of triangle AOB $= \dfrac{1}{2}ab\sin C$

$\qquad\qquad\qquad = \dfrac{1}{2} \times 6 \times 6 \times \sin 2.4619...$

$\qquad\qquad\qquad = 11.3137...\,\text{cm}^2$ ✓

Angle in sector $= 2\pi - (2.4619...)$

$\qquad\qquad\qquad = 3.82126...$ ✓

Area of sector $= \dfrac{1}{2}r^2\,\theta$

$\qquad\qquad\quad = \dfrac{1}{2} \times 6^2 \times 3.82126...$ ✓

$\qquad\qquad\quad = 68.7827...\,\text{cm}^2$

Total area of water $= 80.0964...\,\text{cm}^2$ ✓

$\qquad\qquad\qquad\quad = 80.1\,\text{cm}^2$ (3 s.f.)

(b) 40 cm × 80.0964... cm² = 3203.856... cm³ per second ✓

= 3.203856... litres per second

= 11533.88... litres per hour ✓

≈ 11500 litres per hour (3 s.f.)

(Total for Question 2 is 8 marks)

64

3 (a) A geometric series has first term a and common ratio r.

Prove that the sum, S, of the first n terms of the series is

$$\frac{a(1 - r^n)}{1 - r}$$

(4)

An author will be paid a salary of £27000 for the year 2018. Her contract promises a 4% increase in salary every year, the first increase being given in 2019, so that these annual salaries form a geometric sequence.

(b) Find, to the nearest £100, the salary in the year 2022.

(2)

The author intends to retire in 2040.

(c) Find, to the nearest £1000, the total amount of salary the author will receive in the period from 2018 until she retires at the end of 2040.

(3)

(a) $S = a + ar + \ldots + ar^{n-1}$ ✓

$rS = \qquad ar + ar^2 + \ldots + ar^n$ ✓

$S - rS = a - ar^n$ ✓

$S = \dfrac{a - ar^n}{1 - r}$

$= \dfrac{a(1 - r^n)}{1 - r}$ ✓

(b) $a = 27000, r = 1.04, n = 5$

$ar^{n-1} = 27000(1.04)^4$ ✓

$= £31586.18 \approx £31600$ (to the nearest £100) ✓

(c) 2040 is $n = 23$

$S_{23} = \dfrac{27000(1 - 1.04^{23})}{1 - 1.04}$ ✓

$= £988682.99$ ✓

$= £989000$ to the nearest £1000 ✓

(Total for Question 3 is 9 marks)

Revision Guide
pages 69, 70

LEARN IT!

You need to learn the proof for the sum of a geometric series (and the proof for the sum of an arithmetic series).

Hint Q3b

The nth term of a geometric series is ar^{n-1} where a is the first term, r is the common ratio and n is the term number. The term number for 2022 is $n = 5$.

Problem solving

Series questions will often have a practical application. Geometric series are usually used with financial problems or population growth problems.

Revision Guide
page 87

LEARN IT!

The formula for $\cos(A \pm B)$ is given in the formulae book as is the first line of the proof.

You need to be able to differentiate $\sin x$ and $\cos x$ from first principles.

4   Given that x is measured in radians, prove, from first principles, that the derivative of $\cos x$ is $-\sin x$.

You may assume the formula for $\cos (A \pm B)$ and that as $h \to 0$,

$$\frac{\sin h}{h} \to 1 \text{ and } \frac{\cos h - 1}{h} \to 0$$

(5)

$$f'(x) = \lim_{h \to 0} \frac{f(x + h) - f(x)}{h}$$

$$\frac{d}{dx}\cos x = \lim_{h \to 0} \frac{\cos(x + h) - \cos x}{h} \quad \checkmark$$

$$\cos(x + h) = \cos x \cos h - \sin x \sin h$$

$$\frac{d}{dx}\cos x = \lim_{h \to 0} \frac{\cos x \cos h - \sin x \sin h - \cos x}{h} \quad \checkmark$$

$$= \lim_{h \to 0} \left(\frac{\cos x \cos h - \cos x}{h} + \frac{-\sin x \sin h}{h} \right) \quad \checkmark$$

$$= \lim_{h \to 0} \left(\cos x \frac{(\cos h - 1)}{h} + -\sin x \frac{\sin h}{h} \right)$$

$$= \lim_{h \to 0} \left(\cos x \frac{(\cos h - 1)}{h} + (-\sin x)\frac{\sin h}{h} \right)$$

$$= \lim_{h \to 0} \cos x \left(\frac{(\cos h - 1)}{h} \right) + \lim_{h \to 0} \left((-\sin x)\frac{\sin h}{h} \right) \quad \checkmark$$

$$= \cos x \times 0 + (-\sin x) \times 1$$

$$\frac{d}{dx}\cos x = -\sin x \quad \checkmark$$

(Total for Question 4 is 5 marks)

66

5 $f(x) = 2x^3 - 3x^{\frac{3}{2}} - 4, \quad x > 0$

(a) The root α of the equation $f(x) = 0$ lies in the interval [1.7, 1.8]

Taking 1.8 as a first approximation to α, apply the Newton–Raphson process twice to $f(x)$ to show that $\alpha \approx 1.768$ to 3 decimal places.

(6)

(b) Find the x-coordinate of the stationary point on f(x).

(3)

Revision Guide
page 99

Hint Q5a

The Newton–Raphson formula is given in the formulae book. You can program the formula into your calculator once you are in the examination but not beforehand. If you are not using a program then don't round your answers until the very end. The first step is to differentiate f(x).

(a) $x_{n+1} = x_n - \dfrac{f(x_n)}{f'(x_n)}$

$f'(x) = 6x^2 - \dfrac{9}{2}x^{\frac{1}{2}}$ ✓✓

$x_0 = 1.8$

$x_1 = 1.8 - \dfrac{2(1.8)^3 - 3(1.8)^{\frac{3}{2}} - 4}{6(1.8)^2 - \dfrac{9}{2}(1.8)^{\frac{1}{2}}}$ ✓ $= 1.7687...$ ✓

$x_2 = 1.7687 - \dfrac{(2(1.7687)^3 - 3(1.7687)^{\frac{3}{2}} - 4}{6(1.7687...)^2 - \dfrac{9}{2}(1.7687...)^{\frac{1}{2}}}$ ✓

$= 1.76797002... \approx 1.768$ ✓

(b) $\qquad f'(x) = 0$

$6x^2 - \dfrac{9}{2}x^{\frac{1}{2}} = 0$ ✓

$12x^2 - \dfrac{9}{2}x^{\frac{1}{2}} = 0$

$144x^4 - 81x = 0$ ✓✓

$x = \sqrt[3]{\dfrac{81}{144}}$

$x = 0.825$ (3 s.f.) ✓

Alternative solution

$6x^2 - \dfrac{9}{2}x^{\frac{1}{2}} = 0$ ✓

$1.5x^{\frac{1}{2}}(4x^{\frac{3}{2}} - 3) = 0$

$x^{\frac{3}{2}} = 0.75 \ (x > 0)$ ✓

$x = 0.75^{\frac{2}{3}}$

$x = 0.825$ (3 s.f.) ✓

(Total for Question 5 is 9 marks)

67

Revision Guide
pages 82, 83

Hint

You will need to use the addition formulae and the double angle formulae as well as $\tan\theta = \dfrac{\sin\theta}{\cos\theta}$

Hint

Write $\cos 3\theta$ as $\cos(2\theta + \theta)$ then use $\cos 2\theta = 1 - 2\sin^2\theta$ later in the proof.

Problem solving

To prove an identity, start with one side, then manipulate the expression using known identities until it matches the other side.

6   Prove that

$$\frac{\sin 2\theta \tan\theta}{\cos 3\theta - \cos\theta} \approx \frac{-1}{2\cos\theta}, \ \theta \neq \frac{n\pi}{2}, \quad n \in \mathbb{Z}$$

(5)

Numerator $= \sin 2\theta \tan\theta$

$= 2\sin\theta \cos\theta \dfrac{\sin\theta}{\cos\theta}$

$= 2\sin^2\theta$ ✓

Denominator $= \cos 3\theta - \cos\theta$

$= \cos(2\theta + \theta) - \cos\theta$

$= \cos 2\theta \cos\theta - \sin 2\theta \sin\theta - \cos\theta$ ✓

$= \cos\theta(\cos 2\theta - 1) - \sin 2\theta \sin\theta$

$= \cos\theta(-2\sin^2\theta) - 2\sin\theta \cos\theta \sin\theta$ ✓

$= -2\sin^2\theta \cos\theta - 2\sin^2\theta \cos\theta$

$= -4\sin^2\theta \cos\theta$ ✓

$\dfrac{\text{Numerator}}{\text{Denominator}} = \dfrac{2\sin^2\theta}{-4\sin^2\theta \cos\theta}$

$= \dfrac{-1}{2\cos\theta}$ ✓

(Total for Question 6 is 5 marks)

68

159

7 The rate of change of y with respect to x is proportional to $\ln x$.
   When $y = 0$, $x = 0$ and when $y = 6$, $x = 2$.

   Find y when $x = 8$. Write your answer in the form

   $$\frac{A\ln 2 - B}{\ln 2 - C}$$

   where A, B and C are integers to be found.

   (8)

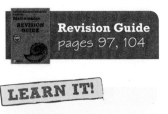

Revision Guide
pages 97, 104

LEARN IT!

$\int k\ln x\,dx$ is a standard integral proof.

Always set $u = \ln x$ when using integration by parts.

Hint

Once the integration is done, substitute in the given values to find c and k. Then substitute $x = 8$ and manipulate into the required form.

$\dfrac{dy}{dx} = k\ln x$ ✓

$\int dy = \int k\ln x\,dx$

$y = kx\ln x - \int k\,dx$ ✓

$= kx\ln x - kx + c$ ✓

When $y = 0$, $x = 0$:

$0 = k(0) \times \ln(0) - k(0) + c$

$c = 0$

So $y = kx\ln x - kx$ ✓

When $y = 6$, $x = 2$:

$6 = k(2) \times \ln(2) - k(2)$

$3 = k\ln 2 - k$

$3 = k(\ln 2 - 1)$ ✓

$k = \dfrac{3}{\ln 2 - 1}$

So $y = \dfrac{3(x\ln x - x)}{\ln 2 - 1}$ ✓

When $x = 8$:

$y = \dfrac{3(8\ln 8 - 8)}{\ln(2) - 1}$ ✓

$= \dfrac{24(\ln 8 - 1)}{\ln 2 - 1}$

$= \dfrac{24(3\ln 2 - 1)}{\ln 2 - 1}$

$= \dfrac{72\ln 2 - 24}{\ln 2 - 1}$ ✓

(Total for Question 7 is 8 marks)

Revision Guide
pages 48, 49, 50

Hint Q8a

Take natural logs of both sides. Don't round your answers until the very end.

Hint Q8b

Use the rules of logarithms, leading to a quadratic equation in y.

LEARN IT!

$a\log b = \log b^a$

$\log a + \log b = \log ab$

$\log a - \log b = \log \dfrac{a}{b}$

8 (a) Solve $3^{2x-1} = 5^{3x+1}$ giving your answer to 3 significant figures.

(3)

(b) Find the values of y such that

$$\log_3(9y-5) + 2\log_3 2 - \log_3 y - \log_3(y+1) = 2, \quad y > \frac{5}{9}, y \in \mathbb{R}$$

(6)

(a) $\quad \ln 3^{2x-1} = \ln 5^{3x+1}$

$\quad (2x-1)\ln 3 = (3x+1)\ln 5 \quad \checkmark$

$\quad 2x\ln 3 - \ln 3 = 3x\ln 5 + \ln 5$

$\quad x(\ln 3^2 - \ln 5^3) = \ln 3 + \ln 5$

$\quad\quad x = \dfrac{\ln 3 + \ln 5}{\ln 9 - \ln 125} \quad \checkmark$

$\quad\quad = -1.029... = -1.03 \text{ (3 s.f.)} \quad \checkmark$

(b) $\log_3(9y-5) + 2\log_3 2 - \log_3 y - \log_3(y+1) = 2$

$\quad \log_3(9y-5) + \log_3 4 - \log_3 y - \log_3(y+1) = 2 \quad \checkmark$

$\quad \log_3 \dfrac{4(9y-5)}{y(y+1)} = 2 \quad \checkmark$

$\quad \text{So} \ \dfrac{4(9y-5)}{y(y+1)} = 3^2 = 9 \quad \checkmark$

$\quad\quad 4(9y-5) = 9y(y+1),$

$\quad\quad 9y^2 - 27y + 20 = 0 \quad \checkmark$

$\quad\quad (3y-4)(3y-5) = 0 \quad \checkmark$

$\quad\quad y = \dfrac{4}{3} \text{ and } y = \dfrac{5}{3} \quad \checkmark$

(Total for Question 8 is 9 marks)

70

161

9 $f(x) = \dfrac{16}{(1-x)^2(3+x)}$

(a) Express $f(x)$ in partial fractions.

(4)

(b) Hence, or otherwise, find the series expansion of $f(x)$, in ascending powers of x, up to and including the term in x^2. Simplify each term.

(8)

(a) $\dfrac{16}{(1-x)^2(3+x)} = \dfrac{A}{1-x} + \dfrac{B}{(1-x)^2} + \dfrac{C}{3+x}$

$16 = A(1-x)(3+x) + B(3+x) + C(1-x)^2$ ✓

Let $x = 1$:

$16 = 4B$

$B = 4$ ✓

Let $x = -3$:

$16 = 16C$

$C = 1$ ✓

Equating coefficients of x^2:

$0 = -A + C$

$A = C = 1$

So, $\dfrac{16}{(1-x)^2(3+x)} = \dfrac{1}{1-x} + \dfrac{4}{(1-x)^2} + \dfrac{1}{3+x}$ ✓

Revision Guide
pages 24, 58

Hint Q9a

There is a repeated factor in the denominator so you must have denominators of $(1-x)$ and $(1-x)^2$.

$f(x) = \dfrac{A}{1-x} + \dfrac{B}{(1-x)^2}$
$\quad + \dfrac{C}{3+x}$

Either use sensible substitutions or equate coefficients to find A, B and C.

Hint Q9b

Always expand one power further than you are asked for in case dividing or multiplying includes terms you weren't expecting. You can always discard them at the end if you don't need them.

Hint Q9b

$(3+x)^{-1}$ is the awkward expansion here as you need to take a common factor of 3 out first to get $3^{-1}(1 + \frac{1}{3}x)^{-1}$.

71

(b) $f'(x) = \dfrac{1}{1-x} + \dfrac{4}{(1-x)^2} + \dfrac{1}{3+x}$

$= (1-x)^{-1} + 4(1-x)^{-2} + (3+x)^{-1}$

$(1-x)^{-1} = 1 + (-1)(-x) + \dfrac{(-1)(-2)(-x)^2}{2!}$
$$+ \dfrac{(-1)(-2)(-3)(-x)^3}{3!} + ... \checkmark$$

$= 1 + x + x^2 + x^3 + ... \checkmark$

$(1-x)^{-2} = 1 + (-2)(-x) + \dfrac{(-2)(-3)(-x)^2}{2!}$
$$+ \dfrac{(-2)(-3)(-4)(-x)^3}{3!} + ... \checkmark$$

$= 1 + 2x + 3x^2 + 4x^3 + ... \checkmark$

$(3+x)^{-1} = (3)^{-1}\left(1 + \dfrac{1}{3}x\right)^{-1} \checkmark$

$= \dfrac{1}{3}\left(1 + (-1)\dfrac{1}{3}x + \dfrac{(-1)(-2)}{2!}\left(\dfrac{1}{3}x\right)^2\right.$
$$\left.+ \dfrac{(-1)(-2)(-3)}{3!}\left(\dfrac{1}{3}x\right)^3 + ...\right) \checkmark$$

$= \dfrac{1}{3}\left(1 - \dfrac{1}{3}x + \dfrac{1}{9}x^2 - \dfrac{1}{27}x^3 + ...\right) \checkmark$

$f(x) = (1 + x + x^2 + x^3 + ...) + 4(1 + 2x + 3x^2 + 4x^3 + ...)$
$$+ \dfrac{1}{3}\left(1 - \dfrac{1}{3}x + \dfrac{1}{9}x^2 - \dfrac{1}{27}x^3 + ...\right)$$

$\approx \dfrac{16}{3} + \dfrac{80}{9}x + \dfrac{352}{27}x^2 \checkmark$

(Total for Question 9 is 12 marks)

10 A curve has equation

$$y^2 + 4y = e^x \sin^2 x \quad \text{for } 0 < x < 2\pi$$

Find the values of x where $\dfrac{dy}{dx} = 0$.

(7)

$2y\dfrac{dy}{dx} + 4\dfrac{dy}{dx}$ ✓ $= e^x \sin^2 x + 2e^x \sin x \cos x$ ✓

$(2y + 4)\dfrac{dy}{dx} = e^x \sin^2 x + 2e^x \sin x \cos x$

$\dfrac{dy}{dx} = \dfrac{e^x \sin^2 x + 2e^x \sin x \cos x}{2y + 4}$ ✓

When $\dfrac{dy}{dx} = 0$, $0 = e^x \sin^2 x + 2e^x \sin x \cos x$

$0 = e^x \sin x(\sin x + 2\cos x)$ ✓

$e^x \neq 0$ so either $\sin x = 0$ or $\sin x + 2\cos x = 0$

$\sin x = 0$ gives $x = \pi$ ✓

$\sin x + 2\cos x = 0$ gives $\tan x = -2$ ✓

So, $x = 2.03$ rad and $x = 5.18$ rad ✓

(Total for Question 10 is 7 marks)

Revision Guide
page 94

Hint

There are three different types of differentiation here. The left-hand side needs to be differentiated implicitly, that is, differentiate with respect to y and then put $\dfrac{dy}{dx}$ next to each term.

The right-hand side is a product rule and then $\sin^2 x$ can be differentiated using a substitution (although it's likely that you will be able to do this in your head).

LEARN IT!

The derirative of e^x is e^x.

Hint

Once the differentiation is complete, substitute in $\dfrac{dy}{dx} = 0$ and solve for x.

LEARN IT!

$e^x \neq 0$ because the graph of $y = e^x$ never crosses the x-axis.

Revision Guide
page 109

Problem solving

Use the chain rule or the quotient rule to get an expression for $\frac{dN}{dt}$ in terms of t.

Hint Q11b

To get $\frac{dN}{dt}$ in terms of N only, you will need to use the original equation and substitute for $8e^{-0.2t}$. The algebraic manipulation is quite tricky.

Problem solving

You are looking for the value, T, that maximises $\frac{dN}{dt}$ so differentiate again and justify your maximum.

11 Scientists are studying the number of bank voles in a woodland. The number, N, at time t months after the start of the study is modelled by the equation

$$N = \frac{660}{3 + 8e^{-0.2t}}, t \geqslant 0, t \in \mathbb{R}$$

(a) Find the number of bank voles at the start of the study.

(1)

(b) Show that the rate of increase, $\frac{dN}{dt}$, is given by

$$\frac{dN}{dt} = \frac{220N - N^2}{1100}$$

(5)

The rate of increase in bank voles is a maximum after T months.

(c) Find the value of T as predicted by this model and state the maximum number of bank voles in the woodland. Justify your answers.

(6)

(a) When $t = 0$, $N = \dfrac{660}{3 + 8} = 60$ ✓

(b) $N = 660(3 + 8e^{-0.2t})^{-1}$

$$\frac{dN}{dt} = -660(3 + 8e^{-0.2t})^{-2}(8(-0.2)e^{-0.2t}) \checkmark$$

$$\frac{dN}{dt} = \frac{660 \times 8e^{-0.2t}}{5(3 + 8e^{-0.2t})^2} \checkmark$$

So $\dfrac{dN}{dt} = \dfrac{660\left(\dfrac{660 - 3N}{N}\right)}{5\left(\dfrac{660}{N}\right)^2}$ ✓

Using the original equation:

$3N + 8Ne^{-0.2t} = 660$

$8e^{-0.2t} = \dfrac{660 - 3N}{N}$ ✓

$$= \frac{1}{5} \times 660 \times \frac{660 - 3N}{N} \times \frac{N^2}{660^2}$$

$$= \frac{660N - 3N^2}{3300}$$

$$= \frac{220N - N^2}{1100} \checkmark$$

(c) To find the maximum value of the rate of increase $\left(\dfrac{dN}{dt}\right)$,

find $\dfrac{d^2 N}{dt^2}$: ✓

$$\dfrac{d^2N}{dt^2} = \dfrac{220 - 2N}{1100}$$

$\dfrac{220 - 2N}{1100} = 0$ when $N = 110$

$N = 110$ corresponds to a maximum ✓ since differentiating

again gives $\dfrac{-2}{1100}$ which is < 0 ✓

When $N = 110$, using the original equation:

$$110 = \dfrac{660}{3 + 8e^{-0.2t}}$$

$3 + 8e^{-0.2t} = 6$

$e^{-0.2t} = \dfrac{3}{8}$ ✓

So, $-0.2t = \ln\left(\dfrac{3}{8}\right)$ ✓

giving $t = 4.904\ldots \approx 4.9$ months ✓

(Total for Question 11 is 12 marks)

Revision Guide
pages 86, 93

Hint Q12a

Differentiate x and y with respect to t.
$$\frac{dy}{dx} = \frac{dy}{dt} \div \frac{dx}{dt}$$

Hint Q12b

Find x, y and $\frac{dy}{dx}$ when $t = 2$ then use $y - y_1 = m(x - x_1)$ to find the equation.

Hint Q12c

Substitute the parametric equations for x and y into the equation of the tangent. Rearrange this to form a cubic equation. You know that $(t - 2)$ is a factor so you can find the other factors by long division or other methods. There is only one other value of t so use this to find x and y.

12 The curve C has parametric equations
$$x = t^3,\ y = t^2 + 2t$$

(a) Find an expression for $\frac{dy}{dx}$ in terms of t.

(2)

(b) Find the cartesian equation of the tangent to C when $t = 2$.

(3)

The tangent intersects the curve C again at the point D.

(c) Find the cartesian coordinates of D.

(6)

(a) $\dfrac{dx}{dt} = 3t^2$ ⎫
    $\dfrac{dy}{dt} = 2t + 2$ ⎬ ✓

    $\dfrac{dy}{dx} = \dfrac{2t + 2}{3t^2}$ ✓

(b) When $t = 2$, $x = 8$ and $y = 8$ ⎫
    When $t = 2$, $\dfrac{dy}{dx} = \dfrac{1}{2}$ ⎬ ✓

    $y - 8 = \dfrac{1}{2}(x - 8)$ ✓

    $y = \dfrac{1}{2}x + 4$ ✓

76

(c) $\qquad t^2 + 2t = \dfrac{1}{2}t^3 + 4$ ✓

$\qquad 2t^2 + 4t = t^3 + 8$

$t^3 - 2t^2 - 4t + 8 = 0$ ✓

$t = 2$ is a known root of the equation so $(t - 2)$ is a

factor of the cubic expression. ✓

$$
\begin{array}{r}
t^2 - 4 \\
t - 2 \overline{) t^3 - 2t^2 - 4t + 8} \\
\underline{t^3 - 2t^2} \\
0 - 4t + 8 \\
\underline{-4t + 8} \\
0
\end{array}
$$

$t^3 - 2t^2 - 4t + 8 = (t - 2)(t^2 - 4)$ ✓

$\qquad = (t - 2)(t - 2)(t + 2)$ ✓

So D corresponds to $t = -2$

When $t = -2$, $x = -8$, $y = 0$

so D has coordinates $(-8, 0)$ ✓

(Total for Question 12 is 11 marks)

TOTAL FOR PAPER IS 100 MARKS

Revision Guide
pages 136, 137

Hint Q1a

You can find PMCCs from given data values using your calculator.

Hint Q1b

This is a one-tailed test. Remember to state your conclusion in the context of the question.

Watch out!

You must state your hypotheses in terms of the population PMCC, p, and show a comparison between the r-value from the table and the r-value you have calculated.

Paper 3: Statistics and Mechanics
SECTION A: STATISTICS

Answer all questions. Write your answers in the spaces provided.

1 A random sample of 10 days in Perth, Western Australia, in September 2015, was chosen, and for each day a mean daily temperature and a mean daily pressure reading was taken.

Mean daily temperature (°C)	11.8	12.3	22.5	11.2	12.1	13.6	17.7	14.2	20.1	15.7
Mean daily pressure (hPa)	1025	1028	1015	1018	1025	1026	1022	1029	1018	1022

(a) Calculate the product moment correlation coefficient for these data.

(1)

(b) Stating your hypotheses clearly, test, at the 2.5% level of significance, whether or not the product moment correlation coefficient for these data is less than zero.

(3)

(c) Give an interpretation of the value 2.5% in your hypothesis test.

(1)

(d) Suggest how you could make better use of the large data set to investigate the relationship between the mean daily temperature and the mean daily pressure.

(1)

(a) $r = -0.6507$ ✓

(b) $H_0: \rho = 0$, $H_1: \rho < 0$

$n = 10$, significance level = 2.5%, $r = 0.6319$

So the critical region is $r < -0.6319$

Since $-0.6507 < -0.6319$ ✓, there is sufficient

evidence to reject H_0 and accept H_1 and conclude that the

product moment correlation coefficient is less than zero. ✓

So there is a significant negative linear correlation between

the mean daily temperature and the mean daily pressure. ✓

(c) The probability of incorrectly rejecting H_0 is 0.025. ✓

(d) Use a larger data sample (or all available data). ✓

(Total for Question 1 is 6 marks)

78

2 A box, *C*, contains 7 counters of which 4 are red and 3 are white.

A box, *D*, contains 6 counters of which 2 are red and 4 are white.

A counter is drawn at random from *C* and placed into *D*. A second counter is drawn at random from *C* and placed into *D*. A third counter is then drawn at random from the counters in *D*.

(a) Draw a tree diagram to show this situation, showing all probabilities.

(4)

(b) Find the probability that a white counter is drawn from *D*.

(2)

(c) Given that 2 of the 3 counters drawn from *D* are the same colour, find the probability that they are '2 red and 1 white'.

(5)

Revision Guide
page 129

Hint Q2c

When you are asked for '2 red and 1 white', this can be in any order, so make sure you consider all possible combinations.

Hint Q2c

$$P(B \mid A) = \frac{P(A \cap B)}{P(A)}$$

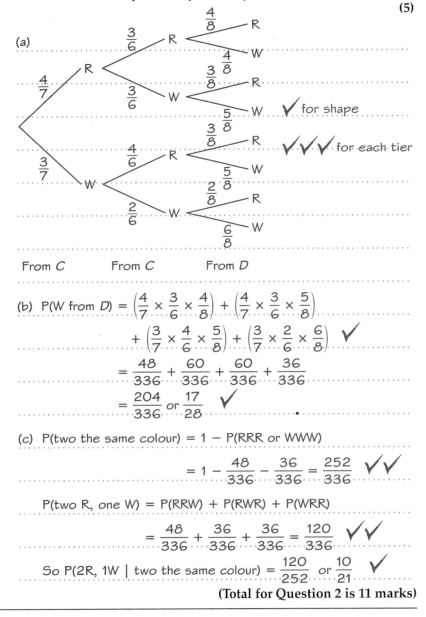

(a)

✔ for shape

✔✔✔ for each tier

From C   From C   From D

(b) P(W from D) $= \left(\frac{4}{7} \times \frac{3}{6} \times \frac{4}{8}\right) + \left(\frac{4}{7} \times \frac{3}{6} \times \frac{5}{8}\right)$

$+ \left(\frac{3}{7} \times \frac{4}{6} \times \frac{5}{8}\right) + \left(\frac{3}{7} \times \frac{2}{6} \times \frac{6}{8}\right)$ ✔

$= \frac{48}{336} + \frac{60}{336} + \frac{60}{336} + \frac{36}{336}$

$= \frac{204}{336}$ or $\frac{17}{28}$ ✔

(c) P(two the same colour) $= 1 - $ P(RRR or WWW)

$= 1 - \frac{48}{336} - \frac{36}{336} = \frac{252}{336}$ ✔✔

P(two R, one W) $=$ P(RRW) $+$ P(RWR) $+$ P(WRR)

$= \frac{48}{336} + \frac{36}{336} + \frac{36}{336} = \frac{120}{336}$ ✔✔

So P(2R, 1W | two the same colour) $= \frac{120}{252}$ or $\frac{10}{21}$ ✔

(Total for Question 2 is 11 marks)

Revision Guide
pages 140, 142, 143

Hint Q3c

Use the inverse normal distribution function on your calculator.

Hint Q3d

Draw a sketch of the distribution. You will see that since $P(Y < 33) = 0.2$, the z-value corresponding to it will be negative. Use the percentage points of the normal distribution table and set up two equations in μ and σ and solve simultaneously.

3 In an aptitude test, the scores, X, are normally distributed with mean 48.5 and standard deviation 9.7

(a) Find $P(X < 52)$

(1)

(b) Find $P(45 < X < 55)$

(1)

(c) A score of k or better was attained by the top 10% of the candidates sitting this aptitude test.

Find the value of k.

(2)

In a second aptitude test, the scores, Y, are normally distributed with mean μ and standard deviation σ.

(d) Given that $P(Y < 33) = 0.2$ and $P(Y < 54) = 0.85$, find the values of μ and σ.

(7)

(a) $P(X < 52) = 0.6409$ ✓

(b) $P(45 < X < 55) = 0.3895$ ✓

(c) $P(X \geqslant k) = 0.1$

   This gives $k = 60.93$ ✓

   so $k = 61$ ✓ (integer score implied)

(d)

$\dfrac{33 - \mu}{\sigma} = -0.8416$ ✓ $\dfrac{54 - \mu}{\sigma} = 1.0364$ ✓

$33 = -0.8416\sigma + \mu$ ①
$54 = 1.0364\sigma + \mu$ ② ✓

Subtract ① from ②:

$21 = 1.8780\sigma$ ✓

So $\sigma = 11.1821...$ ✓

and $\mu = 54 - (1.0364 \times 11.1821...) = 42.41086...$ ✓

So $\mu = 42.4$ (3 s.f.) and $\sigma = 11.2$ (3 s.f.)

(Total for Question 3 is 11 marks)

4   Given that P(A) = 0.45, P(B) = 0.3 and P($B \mid A$) = 0.2, find:

   (a) P($A \cap B$)

   (2)

   (b) P($A' \cap B$)

   (1)

   Event C has P(C) = 0.35 and P($B \mid C$) = 0.4

   Events A and C are mutually exclusive.

   (c) Find P($B \cap C$).

   (2)

   (d) Draw a Venn diagram to illustrate the events A, B and C, giving the probabilities for each region.

   (5)

Revision Guide
pages 127, 128, 138

Hint Q4a

Use the conditional probability definition.

Hint Q4c

Use the conditional probability definition.

Hint Q4d

Don't forget to give the probability for the region outside A, B and C.

(a) P($B \mid A$) = $\dfrac{P(A \cap B)}{P(A)}$

   So 0.2 = $\dfrac{P(A \cap B)}{0.45}$ ✓

   P($A \cap B$) = 0.2 × 0.45 = 0.09 ✓

(b) P($A' \cap B$) = P(B) − P($A \cap B$) = 0.3 − 0.09 = 0.21 ✓

(c) P($B \mid C$) = $\dfrac{P(B \cap C)}{P(C)}$

   So 0.4 = $\dfrac{P(B \cap C)}{0.35}$ ✓

   P($B \cap C$) = 0.4 × 0.35 = 0.14 ✓

(d)

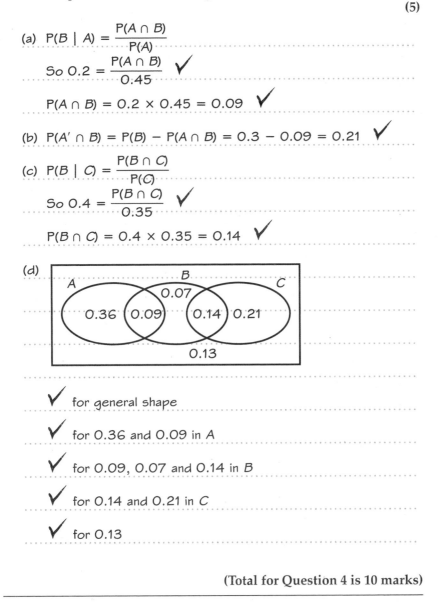

   ✓ for general shape

   ✓ for 0.36 and 0.09 in A

   ✓ for 0.09, 0.07 and 0.14 in B

   ✓ for 0.14 and 0.21 in C

   ✓ for 0.13

(Total for Question 4 is 10 marks)

81

172

Revision Guide
page 145

Problem solving

There are many stages in answering this question. You will need to use the percentage points of the normal distribution table to find the mean thickness for the first machine. Only then can you find the probability in part (a).

Then, think about an appropriate probability distribution to use in part (b).

Hint Q5c

Work out the variance for the sample $\left(\dfrac{\sigma^2}{n}\right)$ before you use a hypothesis test.

Modelling

Always interpret your conclusions in the context of the question.

5   A machine produces metal sheets of thickness T mm, where T is normally distributed with standard deviation 0.15 mm.

Sheets less than 5.5 mm thick or more than 6.0 mm thick **cannot** be used.

Given that 2.5% of sheets have a thickness of less than 5.45 mm:

(a) find the probability than a randomly chosen metal sheet **can** be used.

(5)

Fifteen metal sheets are chosen at random.

(b) Find the probability that fewer than 4 of them **cannot** be used.

(2)

Another machine also produces metal sheets of thickness X mm, where X is normally distributed with standard deviation 0.18 mm.

A random sample of 25 sheets produced by this machine is taken and the sample mean thickness was found to be 5.88 mm.

(c) Stating your hypotheses clearly, and using a 2% level of significance, test whether the mean thickness of metal sheets produced by this machine is greater than 5.8 mm.

(5)

(a)

$P(T < 5.45) = 0.025$ ✓

So $\dfrac{5.45 - \mu}{0.15} = -1.96$ (from % points table) ✓✓

$5.45 - \mu = -1.96 \times 0.15$, giving $\mu = 5.744$ ✓

So, using $\mu = 5.744$ and $\sigma = 0.15$:

P (can be used) $= P(5.5 < T < 6.0) = 0.9042$ ✓

(b) For 15 sheets chosen at random:

Let M represent number of sheets that cannot be used.

$M \sim B(15, 0.0958)$, ✓ where $0.0958 = 1 - 0.9042$

$P(M < 4) = P(M \leqslant 3) = 0.9514$ ✓

82

(c) H_0: $m = 5.8$, H_1: $m > 5.8$ ✓

$\bar{X} \sim N\left(5.8, \left(\dfrac{0.18}{\sqrt{25}}\right)^2\right)$ ✓ and we are testing 5.88

$P(\bar{X} > 5.88) = 0.0131$ ✓

Since $0.0131 < 0.02$, reject H_0 ✓

There is sufficient evidence to support the claim that this

machine produces metal sheets of mean thickness greater

than 5.8 mm. ✓

(Total for Question 5 is 12 marks)

Revision Guide
page 178

Hint

You need to split the time into two parts, 0 to 5π and 5π to 30, and integrate the appropriate expression. Add the answers for the final displacement.

LEARN IT!

Integrate the velocity to find the displacement. If you want to find the displacement after a fixed period of time you can use definite integration.

SECTION B: MECHANICS

Answer **all** questions. Write your answers in the spaces provided.

Unless otherwise indicated, whenever a numerical value of g is required, take $g = 9.8\,\mathrm{m\,s^{-2}}$ and give your answer to either 2 significant figures or 3 significant figures.

6   A particle, P, is moving in a straight line. At time t seconds, $t \geqslant 0$, the velocity of P, $v\,\mathrm{m\,s^{-1}}$, is given by

$$v = \begin{cases} 2 + \sin 0.2t, & 0 < t \leqslant 5\pi \\ 2\cos 0.4t, & 5\pi \leqslant t \leqslant 10\pi \end{cases}$$

Find the displacement of P from its starting position after 30 seconds.

(8)

$$s_1 = \int_0^{5\pi} (2 + \sin 0.2t)\,dt$$

$$= [2t - 5\cos 0.2t]_0^{5\pi} \checkmark\checkmark$$

$$= (10\pi - 5\cos \pi) - (0 - 5\cos 0) \checkmark$$

$$= 10\pi + 5 + 5$$

$$= 10\pi + 10 \checkmark$$

$$s_2 = \int_{5\pi}^{30} (2\cos 0.4t)\,dt$$

$$= [5\sin 0.4t]_{5\pi}^{30} \checkmark\checkmark$$

$$= 5\sin 12 - 5\sin 2\pi$$

$$= 5\sin 12 - 0 \checkmark$$

Displacement after 30 seconds $= 10\pi + 10 + 5\sin 12$

$$= 38.7\,\mathrm{m}\ (3\ \mathrm{s.f.}) \checkmark$$

(Total for Question 6 is 8 marks)

84

7

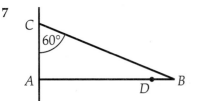

A uniform rod AB, of mass $4m$ and length $6a$, is held in a horizontal position with the end A against a rough vertical wall. One end of a light inextensible string is attached to the rod at B and the other end to a point C on the wall vertically above A. A particle of mass $3m$ is attached to the rod at D, where $AD = 5a$. The angle between the wall and the string at C is $60°$. The rod is in equilibrium in a vertical plane perpendicular to the wall.

(a) Find the tension in the string.

(4)

The coefficient of friction between the rod and the wall is μ. Given that the rod is about to slip,

(b) find the value of μ.

(6)

Revision Guide
pages 172–74

Problem solving

Draw a diagram and mark on it all the forces acting on the rod.

Hint Q7a

Take moments about an appropriate point to find the tension in the string.

Hint Q7b

The rod is about to slip, so friction is limiting. Do not round any answers prematurely. Keep at least 4 decimal places or work with surds.

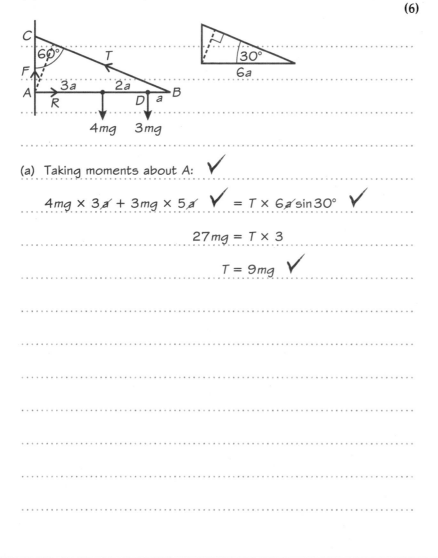

(a) Taking moments about A: ✔

$$4mg \times 3a + 3mg \times 5a \;✔\; = T \times 6a \sin 30° \;✔$$

$$27mg = T \times 3$$

$$T = 9mg \;✔$$

(b) Resolving perpendicular to the wall: ✓

$$R = T \sin 60° = 9mg \times \frac{\sqrt{3}}{2}$$ ✓

The rod is about to slip, so $F = \mu R = \frac{9\sqrt{3}}{2} mg\mu$ ✓

Resolving vertically: ✓

$$F + T \cos 60° = 7mg$$

$$\frac{9\sqrt{3}}{2} mg\mu + \frac{9}{2} mg = 7mg$$ ✓

$$9\sqrt{3}\,\mu = 14 - 9 = 5$$

So $\mu = \dfrac{5}{9\sqrt{3}} = \dfrac{5\sqrt{3}}{27}$ (or 0.3207…) ✓

(Total for Question 7 is 10 marks)

86

8

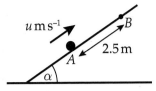

A particle has a mass of 3 kg and is projected from a point A, with a speed of $u\,\mathrm{m\,s^{-1}}$, up a line of greatest slope of a rough inclined plane. The plane makes an angle α with the horizontal where $\tan\alpha = \dfrac{4}{3}$.

The coefficient of friction between the particle and the plane is $\dfrac{1}{6}$.

Given that the particle comes to instantaneous rest at B, where $AB = 2.5\,\mathrm{m}$,

(a) find the value of u.

(7)

The particle then moves back down the inclined plane.

(b) Find the speed of the particle as it passes through A.

(4)

Revision Guide
page 169

Problem solving

Draw a diagram and mark on it all the forces acting on the particle.

Hint Q8a

Use a combination of $F = ma$ and a *suvat* formula to work out u.

Hint Q8b

Now that the particle is moving down the slope, friction will act upwards to oppose the motion.

(a)

$\tan\alpha = \dfrac{4}{3}$

$\sin\alpha = \dfrac{4}{5}$

$\cos\alpha = \dfrac{3}{5}$

Perpendicular to the slope, $R = 3g\cos\alpha = \dfrac{9}{5}g$ ✓

Friction acts down the inclined plane, opposing the motion.

$F = \mu R = \dfrac{1}{6} \times \dfrac{9}{5}g = \dfrac{3}{10}g$ ✓

Component of the weight down the slope $= 3g\sin\alpha$

$= \dfrac{12}{5}g$ ✓

Using $F = ma$, taking up the slope as positive direction:

$-\left(\dfrac{12}{5}g + \dfrac{3}{10}g\right) = 3a$ ✓

$-\dfrac{27}{10}g = 3a$,   so $a = -\dfrac{9}{10}g$ ✓

Using $v^2 = u^2 + 2as$, with $v = 0$, $s = 2.5$ and $a = -\dfrac{9}{10}g$:

$0 = u^2 - 2 \times \dfrac{9}{10}g \times 2.5$ ✓

So, $u^2 = \dfrac{45}{10}g$,   giving $u = 6.64\,\mathrm{m\,s^{-1}}$ (3 s.f.) ✓

87

178

(b)

On the downward journey, friction acts up the slope,

opposing the motion.

Force causing acceleration down the slope = $3g\sin\alpha - F$

$$= \frac{12}{5}g - \frac{3}{10}g$$

$$= \frac{21}{10}g \quad \checkmark$$

Using $F = ma$, taking down the slope as positive direction:

$$\frac{21}{10}g = 3a, \qquad \text{so } a = \frac{7}{10}g \quad \checkmark$$

Using $v^2 = u^2 + 2as$, with $u = 0$, $s = 2.5$ and $a = \frac{7}{10}g$:

$$v^2 = 0 + 2 \times \frac{7}{10}g \times 2.5 = \frac{35}{10}g \quad \checkmark$$

$$v = 5.86\,\text{ms}^{-1} \text{ (3 s.f.)} \quad \checkmark$$

This is the speed of the particle as it passes A.

(Total for Question 8 is 11 marks)

88

9 In this question, **i** and **j** are horizontal unit vectors due east and due north respectively.

At time t seconds, where $t \geqslant 0$, a particle, P, is moving in a horizontal plane with acceleration

$a = [(3t + 2)\mathbf{i} + (t + 4)\mathbf{j}]\,\mathrm{m\,s^{-2}}$

When $t = 2$, the velocity of P is $(6\mathbf{i} + 9\mathbf{j})\,\mathrm{m\,s^{-1}}$.

Find:

(a) the velocity of P at time t seconds

(5)

(b) the speed of P when it is moving on a bearing of 045°.

(5)

Revision Guide
pages 176, 177, 179

Hint Q9a

Use calculus and don't forget to include a constant of integration vector.

Hint Q9b

What will be the connection between the **i** and **j** components of the velocity when P is moving on a bearing of 045°?

You are asked for the speed of P so don't leave your answer in vector form.

(a) $\mathbf{a} = [(3t + 2)\mathbf{i} + (t + 4)\mathbf{j}]\,\mathrm{m\,s^{-2}}$

$\mathbf{v} = \int \mathbf{a}\,dt = \int [(3t + 2)\mathbf{i} + (t + 4)\mathbf{j}]\,dt$ ✓

$= \left(\dfrac{3t^2}{2} + 2t\right)\mathbf{i} + \left(\dfrac{t^2}{2} + 4t\right)\mathbf{j} + \mathbf{c}$ ✓✓

When $t = 2$, $\mathbf{v} = (6\mathbf{i} + 9\mathbf{j})\,\mathrm{m\,s^{-1}}$

So $6\mathbf{i} + 9\mathbf{j} = (6 + 4)\mathbf{i} + (2 + 8)\mathbf{j} + \mathbf{c}$

$6\mathbf{i} + 9\mathbf{j} = 10\mathbf{i} + 10\mathbf{j} + \mathbf{c}$, so $\mathbf{c} = -4\mathbf{i} - \mathbf{j}$ ✓

So $\mathbf{v} = \left(\dfrac{3t^2}{2} + 2t - 4\right)\mathbf{i} + \left(\dfrac{t^2}{2} + 4t - 1\right)\mathbf{j}$ ✓

(b) When P is travelling on a bearing of 045°, the **i** and **j** components of the velocity vector are equal.

So, for **v** above, $\dfrac{3t^2}{2} + 2t - 4 = \dfrac{t^2}{2} + 4t - 1$ ✓

which simplifies to $t^2 - 2t - 3 = 0$

$(t - 3)(t + 1) = 0$

Since $t \geqslant 0$, this gives $t = 3$ ✓

When $t = 3$, $\mathbf{v} = 15.5\mathbf{i} + 15.5\mathbf{j}$ ✓

and the speed of P is $|\mathbf{v}| = \sqrt{15.5^2 + 15.5^2}$ ✓

$= 21.9\,\mathrm{m\,s^{-1}}$ (3 s.f.) ✓

(Total for Question 9 is 10 marks)

89

180

Revision Guide
pages 170, 171

Problem solving

You need to think how to use the given information. As always in projectiles problems, deal with the horizontal motion and vertical motion separately.

Hint Q10a

Use the 15 m in a *suvat* formula to find *u*.

Hint Q10b

Consider the vertical motion from *P* to *R*, then the horizontal motion. What must you calculate to connect both of these?

Modelling

If you have to criticise a model, consider the real-world situation. Make sure you give two different reasons.

10

A ball is projected from a point P which is 42 m above O.

The speed of projection is $u\,\text{ms}^{-1}$ and the angle of projection is 54° to the horizontal.

The highest point of its path, Q, is 15 m above P, and the ball lands on the ground at R.

The ball is modelled as a particle moving freely under gravity in a vertical plane, and the ground is modelled as level.

Taking $g = 10\,\text{ms}^{-2}$, find:

(a) the value of u

(3)

(b) the distance OR.

(6)

(c) State two limitations of this model.

(2)

(a) At Q, using $v^2 = u^2 + 2as$ for the vertical motion (upwards positive):

$0 = (u\sin 54°)^2 - 2 \times 10 \times 15$ ✓

So, $u^2 \sin^2 54° = 300$ ✓

giving $u = 21.409... = 21.4\,\text{ms}^{-1}$ (3 s.f.) ✓

(b) First, find the time for the ball to reach R.

Using $s = ut + \frac{1}{2}at^2$ for the vertical motion (upwards +ve):

$-42 = (21.409 \sin 54°t) - 5t^2$ ✓

So, $5t^2 - (21.409 \sin 54°t) - 42 = 0$ ✓

$5t^2 - 17.32t - 42 = 0$

So, $t = \dfrac{17.32 \pm \sqrt{300 + 840}}{10}$ ✓

giving $t = 5.11\,\text{s}$ ✓

So, OR = horizontal speed × time

$= 21.4 \cos 54° \times 5.11$ ✓

$= 64.3\,\text{m}$ (3 s.f.) ✓

(c) $g = 9.8\,\text{ms}^{-2}$ would give a more accurate result

Air resistance has not been taken into account

Wind speed has not been taken into account ✓✓

Rotational effects have not been taken into account } for any

Ball is unlikely to move in a purely vertical plane two

Ground may not be level

(Total for Question 10 is 11 marks)

TOTAL FOR PAPER IS 100 MARKS

Pure Mathematics

Mensuration

Surface area of sphere = $4\pi r^2$

Area of curved surface of cone = $\pi r \times$ slant height

Arithmetic series

$$S_n = \frac{1}{2}n(a + 1) = \frac{1}{2}n[2a + (n - 1)d]$$

Binomial series

$$(a + b)^n = a^n + \binom{n}{1}a^{n-1}b + \binom{n}{2}a^{n-2}b^2 + \ldots + \binom{n}{r}a^{n-r}b^r + \ldots + b^n \ (n \in \mathbb{N})$$

where $\binom{n}{r} = {}^nC_r = \dfrac{n!}{r!(n - r)!}$

$$(1 + x)^n = 1 + nx + \frac{n(n - 1)}{1 \times 2}x^2 + \ldots + \frac{n(n - 1) \ldots (n - r + 1)}{1 \times 2 \times \ldots \times r}x^r + \ldots \ (|x| < 1, n \in \mathbb{R})$$

Logarithms and exponentials

$$\log_a x = \frac{\log_b x}{\log_b a} \qquad\qquad e^{x\ln a} = a^x$$

Geometric series

$$S_n = \frac{a(1 - r^n)}{1 - r} \qquad\qquad S_\infty = \frac{a}{1 - r} \text{ for } |r| < 1$$

Trigonometric identities

$$\sin(A \pm B) = \sin A\cos B \pm \cos A\sin B$$

$$\cos(A \pm B) = \cos A\cos B \mp \sin A\sin B$$

$$\tan(A \pm B) = \frac{\tan A \pm \tan B}{1 \mp \tan A \tan B} \qquad \left(A \pm B \neq \left(k + \frac{1}{2}\right)\pi\right)$$

$$\sin A + \sin B = 2\sin\frac{A + B}{2}\cos\frac{A - B}{2}$$

$$\sin A - \sin B = 2\cos\frac{A + B}{2}\sin\frac{A - B}{2}$$

$$\cos A + \cos B = 2\cos\frac{A + B}{2}\cos\frac{A - B}{2}$$

$$\cos A - \cos B = -2\sin\frac{A + B}{2}\sin\frac{A - B}{2}$$

Small angle approximations

$$\sin\theta \approx \theta$$

$$\cos\theta \approx 1 - \frac{\theta^2}{2}$$

$$\tan\theta \approx \theta \qquad\qquad \text{where } \theta \text{ is measured in radians}$$

Differentiation

First principles

$$f'(x) = \lim_{h \to 0} \frac{f(x+h) - f(x)}{h}$$

f(x)	f'(x)
$\tan kx$	$k \sec^2 kx$
$\sec kx$	$k \sec kx \tan kx$
$\cot kx$	$-k \operatorname{cosec}^2 kx$
$\operatorname{cosec} kx$	$-k \operatorname{cosec} kx \cot kx$
$\dfrac{f(x)}{g(x)}$	$\dfrac{f'(x)g(x) - f(x)g'(x)}{(g(x))^2}$

Integration (+ constant)

f(x)	$\int f(x)dx$
$\sec^2 kx$	$\dfrac{1}{k}\tan kx$
$\tan kx$	$\dfrac{1}{k}\ln|\sec kx|$
$\cot kx$	$\dfrac{1}{k}\ln|\sin kx|$
$\operatorname{cosec} kx$	$-\dfrac{1}{k}\ln|\operatorname{cosec} kx + \cot kx|,\ \dfrac{1}{k}\ln\left|\tan\left(\dfrac{1}{2}kx\right)\right|$
$\sec kx$	$\dfrac{1}{k}\ln|\sec kx + \tan kx|,\ \dfrac{1}{k}\ln\left|\tan\left(\dfrac{1}{2}kx + \dfrac{1}{4}\pi\right)\right|$

$$\int u\frac{dv}{dx}dx = uv - \int v\frac{du}{dx}dx$$

Numerical methods

The trapezium rule: $\displaystyle\int_a^b y\,dx \approx \frac{1}{2}h\{(y_0 + y_n) + 2(y_1 + y_2 + \ldots + y_{n-1})\}$, where $h = \dfrac{b-a}{n}$

The Newton-Raphson iteration for solving $f(x) = 0$: $x_{n+1} = x_n - \dfrac{f(x_n)}{f'(x_n)}$

Statistics

Probability

$$P(A') = 1 - P(A)$$

$$P(A \cup B) = P(A) + P(B) - P(A \cap B)$$

$$P(A \cap B) = P(A)P(B \mid A)$$

$$P(A \mid B) = \frac{P(B \mid A)P(A)}{P(B \mid A)P(A) + P(B \mid A')P(A')}$$

For independent events A and B:

$$P(B \mid A) = P(B)$$

$$P(A \mid B) = P(A)$$

$$P(A \cap B) = P(A)\,P(B)$$

Standard deviation

Standard deviation = $\sqrt{(\text{variance})}$

Interquartile range = IQR = $Q_3 - Q_1$

For a set of n values $x_1, x_2, \ldots x_i, \ldots x_n$

$$S_{xx} = \Sigma(x_i - \bar{x})^2 = \Sigma x_i^2 - \frac{(\Sigma x_i)^2}{n}$$

Standard deviation = $\sqrt{\dfrac{S_{xx}}{n}}$ or $\sqrt{\dfrac{\Sigma x^2}{n} - \bar{x}^2}$

Discrete distributions

Distribution of X	$P(X = x)$	Mean	Variance
Binomial $B(n, p)$	$\binom{n}{x} p^x (1 - p)^{n-x}$	np	$np(1 - p)$

Sampling distributions

For a random sample of n observations from $N(\mu, \sigma^2)$

$$\frac{\bar{X} - \mu}{\sigma / \sqrt{n}} \sim N(0, 1)$$

Statistical tables

The following statistical tables are provided:

Binomial cumulative distribution function (page 186)

Percentage points of the normal distribution (page 191)

Critical values for correlation coefficients (page 192)

Mechanics

Kinematics

For motion in a straight line with constant acceleration:

$v = u + at$

$s = ut + \dfrac{1}{2}at^2$

$s = vt - \dfrac{1}{2}at^2$

$v^2 = u^2 + 2as$

$s = \dfrac{1}{2}(u + v)t$

Binomial cumulative distribution function

The tabulated value is $P(X \leqslant x)$, where X has a binomial distribution with index n and parameter p.

$p =$	0.05	0.10	0.15	0.20	0.25	0.30	0.35	0.40	0.45	0.50
$n = 5, x = 0$	0.7738	0.5905	0.4437	0.3277	0.2373	0.1681	0.1160	0.0778	0.0503	0.0312
1	0.9774	0.9185	0.8352	0.7373	0.6328	0.5282	0.4284	0.3370	0.2562	0.1875
2	0.9988	0.9914	0.9734	0.9421	0.8965	0.8369	0.7648	0.6826	0.5931	0.5000
3	1.0000	0.9995	0.9978	0.9933	0.9844	0.9692	0.9460	0.9130	0.8688	0.8125
4	1.0000	1.0000	0.9999	0.9997	0.9990	0.9976	0.9947	0.9898	0.9815	0.9688
$n = 6, x = 0$	0.7351	0.5314	0.3771	0.2621	0.1780	0.1176	0.0754	0.0467	0.0277	0.0156
1	0.9672	0.8857	0.7765	0.6554	0.5339	0.4202	0.3191	0.2333	0.1636	0.1094
2	0.9978	0.9842	0.9527	0.9011	0.8306	0.7443	0.6471	0.5443	0.4415	0.3438
3	0.9999	0.9987	0.9941	0.9830	0.9624	0.9295	0.8826	0.8208	0.7447	0.6563
4	1.0000	0.9999	0.9996	0.9984	0.9954	0.9891	0.9777	0.9590	0.9308	0.8906
5	1.0000	1.0000	1.0000	0.9999	0.9998	0.9993	0.9982	0.9959	0.9917	0.9844
$n = 7, x = 0$	0.6983	0.4783	0.3206	0.2097	0.1335	0.0824	0.0490	0.0280	0.0152	0.0078
1	0.9556	0.8503	0.7166	0.5767	0.4449	0.3294	0.2338	0.1586	0.1024	0.0625
2	0.9962	0.9743	0.9262	0.8520	0.7564	0.6471	0.5323	0.4199	0.3164	0.2266
3	0.9998	0.9973	0.9879	0.9667	0.9294	0.8740	0.8002	0.7102	0.6083	0.5000
4	1.0000	0.9998	0.9988	0.9953	0.9871	0.9712	0.9444	0.9037	0.8471	0.7734
5	1.0000	1.0000	0.9999	0.9996	0.9987	0.9962	0.9910	0.9812	0.9643	0.9375
6	1.0000	1.0000	1.0000	1.0000	0.9999	0.9998	0.9994	0.9984	0.9963	0.9922
$n = 8, x = 0$	0.6634	0.4305	0.2725	0.1678	0.1001	0.0576	0.0319	0.0168	0.0084	0.0039
1	0.9428	0.8131	0.6572	0.5033	0.3671	0.2553	0.1691	0.1064	0.0632	0.0352
2	0.9942	0.9619	0.8948	0.7969	0.6785	0.5518	0.4278	0.3154	0.2201	0.1445
3	0.9996	0.9950	0.9786	0.9437	0.8862	0.8059	0.7064	0.5941	0.4770	0.3633
4	1.0000	0.9996	0.9971	0.9896	0.9727	0.9420	0.8939	0.8263	0.7396	0.6367
5	1.0000	1.0000	0.9998	0.9988	0.9958	0.9887	0.9747	0.9502	0.9115	0.8555
6	1.0000	1.0000	1.0000	0.9999	0.9996	0.9987	0.9964	0.9915	0.9819	0.9648
7	1.0000	1.0000	1.0000	1.0000	1.0000	0.9999	0.9998	0.9993	0.9983	0.9961
$n = 9, x = 0$	0.6302	0.3874	0.2316	0.1342	0.0751	0.0404	0.0207	0.0101	0.0046	0.0020
1	0.9288	0.7748	0.5995	0.4362	0.3003	0.1960	0.1211	0.0705	0.0385	0.0195
2	0.9916	0.9470	0.8591	0.7382	0.6007	0.4628	0.3373	0.2318	0.1495	0.0898
3	0.9994	0.9917	0.9661	0.9144	0.8343	0.7297	0.6089	0.4826	0.3614	0.2539
4	1.0000	0.9991	0.9944	0.9804	0.9511	0.9012	0.8283	0.7334	0.6214	0.5000
5	1.0000	0.9999	0.9994	0.9969	0.9900	0.9747	0.9464	0.9006	0.8342	0.7461
6	1.0000	1.0000	1.0000	0.9997	0.9987	0.9957	0.9888	0.9750	0.9502	0.9102
7	1.0000	1.0000	1.0000	1.0000	0.9999	0.9996	0.9986	0.9962	0.9909	0.9805
8	1.0000	1.0000	1.0000	1.0000	1.0000	1.0000	0.9999	0.9997	0.9992	0.9980
$n = 10, x = 0$	0.5987	0.3487	0.1969	0.1074	0.0563	0.0282	0.0135	0.0060	0.0025	0.0010
1	0.9139	0.7361	0.5443	0.3758	0.2440	0.1493	0.0860	0.0464	0.0233	0.0107
2	0.9885	0.9298	0.8202	0.6778	0.5256	0.3828	0.2616	0.1673	0.0996	0.0547
3	0.9990	0.9872	0.9500	0.8791	0.7759	0.6496	0.5138	0.3823	0.2660	0.1719
4	0.9999	0.9984	0.9901	0.9672	0.9219	0.8497	0.7515	0.6331	0.5044	0.3770
5	1.0000	0.9999	0.9986	0.9936	0.9803	0.9527	0.9051	0.8338	0.7384	0.6230
6	1.0000	1.0000	0.9999	0.9991	0.9965	0.9894	0.9740	0.9452	0.8980	0.8281
7	1.0000	1.0000	1.0000	0.9999	0.9996	0.9984	0.9952	0.9877	0.9726	0.9453
8	1.0000	1.0000	1.0000	1.0000	1.0000	0.9999	0.9995	0.9983	0.9955	0.9893
9	1.0000	1.0000	1.0000	1.0000	1.0000	1.0000	1.0000	0.9999	0.9997	0.9990

$p =$	0.05	0.10	0.15	0.20	0.25	0.30	0.35	0.40	0.45	0.50
$n = 12, x = 0$	0.5404	0.2824	0.1422	0.0687	0.0317	0.0138	0.0057	0.0022	0.0008	0.0002
1	0.8816	0.6590	0.4435	0.2749	0.1584	0.0850	0.0424	0.0196	0.0083	0.0032
2	0.9804	0.8891	0.7358	0.5583	0.3907	0.2528	0.1513	0.0834	0.0421	0.0193
3	0.9978	0.9744	0.9078	0.7946	0.6488	0.4925	0.3467	0.2253	0.1345	0.0730
4	0.9998	0.9957	0.9761	0.9274	0.8424	0.7237	0.5833	0.4382	0.3044	0.1938
5	1.0000	0.9995	0.9954	0.9806	0.9456	0.8822	0.7873	0.6652	0.5269	0.3872
6	1.0000	0.9999	0.9993	0.9961	0.9857	0.9614	0.9154	0.8418	0.7393	0.6128
7	1.0000	1.0000	0.9999	0.9994	0.9972	0.9905	0.9745	0.9427	0.8883	0.8062
8	1.0000	1.0000	1.0000	0.9999	0.9996	0.9983	0.9944	0.9847	0.9644	0.9270
9	1.0000	1.0000	1.0000	1.0000	1.0000	0.9998	0.9992	0.9972	0.9921	0.9807
10	1.0000	1.0000	1.0000	1.0000	1.0000	1.0000	0.9999	0.9997	0.9989	0.9968
11	1.0000	1.0000	1.0000	1.0000	1.0000	1.0000	1.0000	1.0000	0.9999	0.9998
$n = 15, x = 0$	0.4633	0.2059	0.0874	0.0352	0.0134	0.0047	0.0016	0.0005	0.0001	0.0000
1	0.8290	0.5490	0.3186	0.1671	0.0802	0.0353	0.0142	0.0052	0.0017	0.0005
2	0.9638	0.8159	0.6042	0.3980	0.2361	0.1268	0.0617	0.0271	0.0107	0.0037
3	0.9945	0.9444	0.8227	0.6482	0.4613	0.2969	0.1727	0.0905	0.0424	0.0176
4	0.9994	0.9873	0.9383	0.8358	0.6865	0.5155	0.3519	0.2173	0.1204	0.0592
5	0.9999	0.9978	0.9832	0.9389	0.8516	0.7216	0.5643	0.4032	0.2608	0.1509
6	1.0000	0.9997	0.9964	0.9819	0.9434	0.8689	0.7548	0.6098	0.4522	0.3036
7	1.0000	1.0000	0.9994	0.9958	0.9827	0.9500	0.8868	0.7869	0.6535	0.5000
8	1.0000	1.0000	0.9999	0.9992	0.9958	0.9848	0.9578	0.9050	0.8182	0.6964
9	1.0000	1.0000	1.0000	0.9999	0.9992	0.9963	0.9876	0.9662	0.9231	0.8491
10	1.0000	1.0000	1.0000	1.0000	0.9999	0.9993	0.9972	0.9907	0.9745	0.9408
11	1.0000	1.0000	1.0000	1.0000	1.0000	0.9999	0.9995	0.9981	0.9937	0.9824
12	1.0000	1.0000	1.0000	1.0000	1.0000	1.0000	0.9999	0.9997	0.9989	0.9963
13	1.0000	1.0000	1.0000	1.0000	1.0000	1.0000	1.0000	1.0000	0.9999	0.9995
14	1.0000	1.0000	1.0000	1.0000	1.0000	1.0000	1.0000	1.0000	1.0000	1.0000
$n = 20, x = 0$	0.3585	0.1216	0.0388	0.0115	0.0032	0.0008	0.0002	0.0000	0.0000	0.0000
1	0.7358	0.3917	0.1756	0.0692	0.0243	0.0076	0.0021	0.0005	0.0001	0.0000
2	0.9245	0.6769	0.4049	0.2061	0.0913	0.0355	0.0121	0.0036	0.0009	0.0002
3	0.9841	0.8670	0.6477	0.4114	0.2252	0.1071	0.0444	0.0160	0.0049	0.0013
4	0.9974	0.9568	0.8298	0.6296	0.4148	0.2375	0.1182	0.0510	0.0189	0.0059
5	0.9997	0.9887	0.9327	0.8042	0.6172	0.4164	0.2454	0.1256	0.0553	0.0207
6	1.0000	0.9976	0.9781	0.9133	0.7858	0.6080	0.4166	0.2500	0.1299	0.0577
7	1.0000	0.9996	0.9941	0.9679	0.8982	0.7723	0.6010	0.4159	0.2520	0.1316
8	1.0000	0.9999	0.9987	0.9900	0.9591	0.8867	0.7624	0.5956	0.4143	0.2517
9	1.0000	1.0000	0.9998	0.9974	0.9861	0.9520	0.8782	0.7553	0.5914	0.4119
10	1.0000	1.0000	1.0000	0.9994	0.9961	0.9829	0.9468	0.8725	0.7507	0.5881
11	1.0000	1.0000	1.0000	0.9999	0.9991	0.9949	0.9804	0.9435	0.8692	0.7483
12	1.0000	1.0000	1.0000	1.0000	0.9998	0.9987	0.9940	0.9790	0.9420	0.8684
13	1.0000	1.0000	1.0000	1.0000	1.0000	0.9997	0.9985	0.9935	0.9786	0.9423
14	1.0000	1.0000	1.0000	1.0000	1.0000	1.0000	0.9997	0.9984	0.9936	0.9793
15	1.0000	1.0000	1.0000	1.0000	1.0000	1.0000	1.0000	0.9997	0.9985	0.9941
16	1.0000	1.0000	1.0000	1.0000	1.0000	1.0000	1.0000	1.0000	0.9997	0.9987
17	1.0000	1.0000	1.0000	1.0000	1.0000	1.0000	1.0000	1.0000	1.0000	0.9998
18	1.0000	1.0000	1.0000	1.0000	1.0000	1.0000	1.0000	1.0000	1.0000	1.0000

$p =$	0.05	0.10	0.15	0.20	0.25	0.30	0.35	0.40	0.45	0.50
$n = 25, x = 0$	0.2774	0.0718	0.0172	0.0038	0.0008	0.0001	0.0000	0.0000	0.0000	0.0000
1	0.6424	0.2712	0.0931	0.0274	0.0070	0.0016	0.0003	0.0001	0.0000	0.0000
2	0.8729	0.5371	0.2537	0.0982	0.0321	0.0090	0.0021	0.0004	0.0001	0.0000
3	0.9659	0.7636	0.4711	0.2340	0.0962	0.0332	0.0097	0.0024	0.0005	0.0001
4	0.9928	0.9020	0.6821	0.4207	0.2137	0.0905	0.0320	0.0095	0.0023	0.0005
5	0.9988	0.9666	0.8385	0.6167	0.3783	0.1935	0.0826	0.0294	0.0086	0.0020
6	0.9998	0.9905	0.9305	0.7800	0.5611	0.3407	0.1734	0.0736	0.0258	0.0073
7	1.0000	0.9977	0.9745	0.8909	0.7265	0.5118	0.3061	0.1536	0.0639	0.0216
8	1.0000	0.9995	0.9920	0.9532	0.8506	0.6769	0.4668	0.2735	0.1340	0.0539
9	1.0000	0.9999	0.9979	0.9827	0.9287	0.8106	0.6303	0.4246	0.2424	0.1148
10	1.0000	1.0000	0.9995	0.9944	0.9703	0.9022	0.7712	0.5858	0.3843	0.2122
11	1.0000	1.0000	0.9999	0.9985	0.9893	0.9558	0.8746	0.7323	0.5426	0.3450
12	1.0000	1.0000	1.0000	0.9996	0.9966	0.9825	0.9396	0.8462	0.6937	0.5000
13	1.0000	1.0000	1.0000	0.9999	0.9991	0.9940	0.9745	0.9222	0.8173	0.6550
14	1.0000	1.0000	1.0000	1.0000	0.9998	0.9982	0.9907	0.9656	0.9040	0.7878
15	1.0000	1.0000	1.0000	1.0000	1.0000	0.9995	0.9971	0.9868	0.9560	0.8852
16	1.0000	1.0000	1.0000	1.0000	1.0000	0.9999	0.9992	0.9957	0.9826	0.9461
17	1.0000	1.0000	1.0000	1.0000	1.0000	1.0000	0.9998	0.9988	0.9942	0.9784
18	1.0000	1.0000	1.0000	1.0000	1.0000	1.0000	1.0000	0.9997	0.9984	0.9927
19	1.0000	1.0000	1.0000	1.0000	1.0000	1.0000	1.0000	0.9999	0.9996	0.9980
20	1.0000	1.0000	1.0000	1.0000	1.0000	1.0000	1.0000	1.0000	0.9999	0.9995
21	1.0000	1.0000	1.0000	1.0000	1.0000	1.0000	1.0000	1.0000	1.0000	0.9999
22	1.0000	1.0000	1.0000	1.0000	1.0000	1.0000	1.0000	1.0000	1.0000	1.0000
$n = 30, x = 0$	0.2146	0.0424	0.0076	0.0012	0.0002	0.0000	0.0000	0.0000	0.0000	0.0000
1	0.5535	0.1837	0.0480	0.0105	0.0020	0.0003	0.0000	0.0000	0.0000	0.0000
2	0.8122	0.4114	0.1514	0.0442	0.0106	0.0021	0.0003	0.0000	0.0000	0.0000
3	0.9392	0.6474	0.3217	0.1227	0.0374	0.0093	0.0019	0.0003	0.0000	0.0000
4	0.9844	0.8245	0.5245	0.2552	0.0979	0.0302	0.0075	0.0015	0.0002	0.0000
5	0.9967	0.9268	0.7106	0.4275	0.2026	0.0766	0.0233	0.0057	0.0011	0.0002
6	0.9994	0.9742	0.8474	0.6070	0.3481	0.1595	0.0586	0.0172	0.0040	0.0007
7	0.9999	0.9922	0.9302	0.7608	0.5143	0.2814	0.1238	0.0435	0.0121	0.0026
8	1.0000	0.9980	0.9722	0.8713	0.6736	0.4315	0.2247	0.0940	0.0312	0.0081
9	1.0000	0.9995	0.9903	0.9389	0.8034	0.5888	0.3575	0.1763	0.0694	0.0214
10	1.0000	0.9999	0.9971	0.9744	0.8943	0.7304	0.5078	0.2915	0.1350	0.0494
11	1.0000	1.0000	0.9992	0.9905	0.9493	0.8407	0.6548	0.4311	0.2327	0.1002
12	1.0000	1.0000	0.9998	0.9969	0.9784	0.9155	0.7802	0.5785	0.3592	0.1808
13	1.0000	1.0000	1.0000	0.9991	0.9918	0.9599	0.8737	0.7145	0.5025	0.2923
14	1.0000	1.0000	1.0000	0.9998	0.9973	0.9831	0.9348	0.8246	0.6448	0.4278
15	1.0000	1.0000	1.0000	0.9999	0.9992	0.9936	0.9699	0.9029	0.7691	0.5722
16	1.0000	1.0000	1.0000	1.0000	0.9998	0.9979	0.9876	0.9519	0.8644	0.7077
17	1.0000	1.0000	1.0000	1.0000	0.9999	0.9994	0.9955	0.9788	0.9286	0.8192
18	1.0000	1.0000	1.0000	1.0000	1.0000	0.9998	0.9986	0.9917	0.9666	0.8998
19	1.0000	1.0000	1.0000	1.0000	1.0000	1.0000	0.9996	0.9971	0.9862	0.9506
20	1.0000	1.0000	1.0000	1.0000	1.0000	1.0000	0.9999	0.9991	0.9950	0.9786
21	1.0000	1.0000	1.0000	1.0000	1.0000	1.0000	1.0000	0.9998	0.9984	0.9919
22	1.0000	1.0000	1.0000	1.0000	1.0000	1.0000	1.0000	1.0000	0.9996	0.9974
23	1.0000	1.0000	1.0000	1.0000	1.0000	1.0000	1.0000	1.0000	0.9999	0.9993
24	1.0000	1.0000	1.0000	1.0000	1.0000	1.0000	1.0000	1.0000	1.0000	0.9998
25	1.0000	1.0000	1.0000	1.0000	1.0000	1.0000	1.0000	1.0000	1.0000	1.0000

$p =$	0.05	0.10	0.15	0.20	0.25	0.30	0.35	0.40	0.45	0.50
$n = 40, x = 0$	0.1285	0.0148	0.0015	0.0001	0.0000	0.0000	0.0000	0.0000	0.0000	0.0000
1	0.3991	0.0805	0.0121	0.0015	0.0001	0.0000	0.0000	0.0000	0.0000	0.0000
2	0.6767	0.2228	0.0486	0.0079	0.0010	0.0001	0.0000	0.0000	0.0000	0.0000
3	0.8619	0.4231	0.1302	0.0285	0.0047	0.0006	0.0001	0.0000	0.0000	0.0000
4	0.9520	0.6290	0.2633	0.0759	0.0160	0.0026	0.0003	0.0000	0.0000	0.0000
5	0.9861	0.7937	0.4325	0.1613	0.0433	0.0086	0.0013	0.0001	0.0000	0.0000
6	0.9966	0.9005	0.6067	0.2859	0.0962	0.0238	0.0044	0.0006	0.0001	0.0000
7	0.9993	0.9581	0.7559	0.4371	0.1820	0.0553	0.0124	0.0021	0.0002	0.0000
8	0.9999	0.9845	0.8646	0.5931	0.2998	0.1110	0.0303	0.0061	0.0009	0.0001
9	1.0000	0.9949	0.9328	0.7318	0.4395	0.1959	0.0644	0.0156	0.0027	0.0003
10	1.0000	0.9985	0.9701	0.8392	0.5839	0.3087	0.1215	0.0352	0.0074	0.0011
11	1.0000	0.9996	0.9880	0.9125	0.7151	0.4406	0.2053	0.0709	0.0179	0.0032
12	1.0000	0.9999	0.9957	0.9568	0.8209	0.5772	0.3143	0.1285	0.0386	0.0083
13	1.0000	1.0000	0.9986	0.9806	0.8968	0.7032	0.4408	0.2112	0.0751	0.0192
14	1.0000	1.0000	0.9996	0.9921	0.9456	0.8074	0.5721	0.3174	0.1326	0.0403
15	1.0000	1.0000	0.9999	0.9971	0.9738	0.8849	0.6946	0.4402	0.2142	0.0769
16	1.0000	1.0000	1.0000	0.9990	0.9884	0.9367	0.7978	0.5681	0.3185	0.1341
17	1.0000	1.0000	1.0000	0.9997	0.9953	0.9680	0.8761	0.6885	0.4391	0.2148
18	1.0000	1.0000	1.0000	0.9999	0.9983	0.9852	0.9301	0.7911	0.5651	0.3179
19	1.0000	1.0000	1.0000	1.0000	0.9994	0.9937	0.9637	0.8702	0.6844	0.4373
20	1.0000	1.0000	1.0000	1.0000	0.9998	0.9976	0.9827	0.9256	0.7870	0.5627
21	1.0000	1.0000	1.0000	1.0000	1.0000	0.9991	0.9925	0.9608	0.8669	0.6821
22	1.0000	1.0000	1.0000	1.0000	1.0000	0.9997	0.9970	0.9811	0.9233	0.7852
23	1.0000	1.0000	1.0000	1.0000	1.0000	0.9999	0.9989	0.9917	0.9595	0.8659
24	1.0000	1.0000	1.0000	1.0000	1.0000	1.0000	0.9996	0.9966	0.9804	0.9231
25	1.0000	1.0000	1.0000	1.0000	1.0000	1.0000	0.9999	0.9988	0.9914	0.9597
26	1.0000	1.0000	1.0000	1.0000	1.0000	1.0000	1.0000	0.9996	0.9966	0.9808
27	1.0000	1.0000	1.0000	1.0000	1.0000	1.0000	1.0000	0.9999	0.9988	0.9917
28	1.0000	1.0000	1.0000	1.0000	1.0000	1.0000	1.0000	1.0000	0.9996	0.9968
29	1.0000	1.0000	1.0000	1.0000	1.0000	1.0000	1.0000	1.0000	0.9999	0.9989
30	1.0000	1.0000	1.0000	1.0000	1.0000	1.0000	1.0000	1.0000	1.0000	0.9997
31	1.0000	1.0000	1.0000	1.0000	1.0000	1.0000	1.0000	1.0000	1.0000	0.9999
32	1.0000	1.0000	1.0000	1.0000	1.0000	1.0000	1.0000	1.0000	1.0000	1.0000

$p =$	0.05	0.10	0.15	0.20	0.25	0.30	0.35	0.40	0.45	0.50
$n = 50, x = 0$	0.0769	0.0052	0.0003	0.0000	0.0000	0.0000	0.0000	0.0000	0.0000	0.0000
1	0.2794	0.0338	0.0029	0.0002	0.0000	0.0000	0.0000	0.0000	0.0000	0.0000
2	0.5405	0.1117	0.0142	0.0013	0.0001	0.0000	0.0000	0.0000	0.0000	0.0000
3	0.7604	0.2503	0.0460	0.0057	0.0005	0.0000	0.0000	0.0000	0.0000	0.0000
4	0.8964	0.4312	0.1121	0.0185	0.0021	0.0002	0.0000	0.0000	0.0000	0.0000
5	0.9622	0.6161	0.2194	0.0480	0.0070	0.0007	0.0001	0.0000	0.0000	0.0000
6	0.9882	0.7702	0.3613	0.1034	0.0194	0.0025	0.0002	0.0000	0.0000	0.0000
7	0.9968	0.8779	0.5188	0.1904	0.0453	0.0073	0.0008	0.0001	0.0000	0.0000
8	0.9992	0.9421	0.6681	0.3073	0.0916	0.0183	0.0025	0.0002	0.0000	0.0000
9	0.9998	0.9755	0.7911	0.4437	0.1637	0.0402	0.0067	0.0008	0.0001	0.0000
10	1.0000	0.9906	0.8801	0.5836	0.2622	0.0789	0.0160	0.0022	0.0002	0.0000
11	1.0000	0.9968	0.9372	0.7107	0.3816	0.1390	0.0342	0.0057	0.0006	0.0000
12	1.0000	0.9990	0.9699	0.8139	0.5110	0.2229	0.0661	0.0133	0.0018	0.0002
13	1.0000	0.9997	0.9868	0.8894	0.6370	0.3279	0.1163	0.0280	0.0045	0.0005
14	1.0000	0.9999	0.9947	0.9393	0.7481	0.4468	0.1878	0.0540	0.0104	0.0013
15	1.0000	1.0000	0.9981	0.9692	0.8369	0.5692	0.2801	0.0955	0.0220	0.0033
16	1.0000	1.0000	0.9993	0.9856	0.9017	0.6839	0.3889	0.1561	0.0427	0.0077
17	1.0000	1.0000	0.9998	0.9937	0.9449	0.7822	0.5060	0.2369	0.0765	0.0164
18	1.0000	1.0000	0.9999	0.9975	0.9713	0.8594	0.6216	0.3356	0.1273	0.0325
19	1.0000	1.0000	1.0000	0.9991	0.9861	0.9152	0.7264	0.4465	0.1974	0.0595
20	1.0000	1.0000	1.0000	0.9997	0.9937	0.9522	0.8139	0.5610	0.2862	0.1013
21	1.0000	1.0000	1.0000	0.9999	0.9974	0.9749	0.8813	0.6701	0.3900	0.1611
22	1.0000	1.0000	1.0000	1.0000	0.9990	0.9877	0.9290	0.7660	0.5019	0.2399
23	1.0000	1.0000	1.0000	1.0000	0.9996	0.9944	0.9604	0.8438	0.6134	0.3359
24	1.0000	1.0000	1.0000	1.0000	0.9999	0.9976	0.9793	0.9022	0.7160	0.4439
25	1.0000	1.0000	1.0000	1.0000	1.0000	0.9991	0.9900	0.9427	0.8034	0.5561
26	1.0000	1.0000	1.0000	1.0000	1.0000	0.9997	0.9955	0.9686	0.8721	0.6641
27	1.0000	1.0000	1.0000	1.0000	1.0000	0.9999	0.9981	0.9840	0.9220	0.7601
28	1.0000	1.0000	1.0000	1.0000	1.0000	1.0000	0.9993	0.9924	0.9556	0.8389
29	1.0000	1.0000	1.0000	1.0000	1.0000	1.0000	0.9997	0.9966	0.9765	0.8987
30	1.0000	1.0000	1.0000	1.0000	1.0000	1.0000	0.9999	0.9986	0.9884	0.9405
31	1.0000	1.0000	1.0000	1.0000	1.0000	1.0000	1.0000	0.9995	0.9947	0.9675
32	1.0000	1.0000	1.0000	1.0000	1.0000	1.0000	1.0000	0.9998	0.9978	0.9836
33	1.0000	1.0000	1.0000	1.0000	1.0000	1.0000	1.0000	0.9999	0.9991	0.9923
34	1.0000	1.0000	1.0000	1.0000	1.0000	1.0000	1.0000	1.0000	0.9997	0.9967
35	1.0000	1.0000	1.0000	1.0000	1.0000	1.0000	1.0000	1.0000	0.9999	0.9987
36	1.0000	1.0000	1.0000	1.0000	1.0000	1.0000	1.0000	1.0000	1.0000	0.9995
37	1.0000	1.0000	1.0000	1.0000	1.0000	1.0000	1.0000	1.0000	1.0000	0.9998
38	1.0000	1.0000	1.0000	1.0000	1.0000	1.0000	1.0000	1.0000	1.0000	1.0000

Percentage points of the normal distribution

The values z in the table are those which a random variable $Z - N(0, 1)$ exceeds with probability p; that is,

$$P(Z > z) = 1 - \Phi(z) = p.$$

p	z	p	z
0.5000	0.0000	0.0500	1.6449
0.4000	0.2533	0.0250	1.9600
0.3000	0.5244	0.0100	2.3263
0.2000	0.8416	0.0050	2.5758
0.1500	1.0364	0.0010	3.0902
0.1000	1.2816	0.0005	3.2905

Critical values for correlation coefficients

These tables concern tests of the hypothesis that a population correlation coefficient ρ is 0. The values in the tables are the minimum values which need to be reached by a sample correlation coefficient in order to be significant at the level shown, on a one-tailed test.

Product moment coefficient					Sample	Spearman's coefficient		
Level					size, n	Level		
0.10	0.05	0.025	0.01	0.005		0.05	0.025	0.01
0.8000	0.9000	0.9500	0.9800	0.9900	4	1.0000	–	–
0.6870	0.8054	0.8783	0.9343	0.9587	5	0.9000	1.0000	1.0000
0.6084	0.7293	0.8114	0.8822	0.9172	6	0.8286	0.8857	0.9429
0.5509	0.6694	0.7545	0.8329	0.8745	7	0.7143	0.7857	0.8929
0.5067	0.6215	0.7067	0.7887	0.8343	8	0.6429	0.7381	0.8333
0.4716	0.5822	0.6664	0.7498	0.7977	9	0.6000	0.7000	0.7833
0.4428	0.5494	0.6319	0.7155	0.7646	10	0.5636	0.6485	0.7455
0.4187	0.5214	0.6021	0.6851	0.7348	11	0.5364	0.6182	0.7091
0.3981	0.4973	0.5760	0.6581	0.7079	12	0.5035	0.5874	0.6783
0.3802	0.4762	0.5529	0.6339	0.6835	13	0.4835	0.5604	0.6484
0.3646	0.4575	0.5324	0.6120	0.6614	14	0.4637	0.5385	0.6264
0.3507	0.4409	0.5140	0.5923	0.6411	15	0.4464	0.5214	0.6036
0.3383	0.4259	0.4973	0.5742	0.6226	16	0.4294	0.5029	0.5824
0.3271	0.4124	0.4821	0.5577	0.6055	17	0.4142	0.4877	0.5662
0.3170	0.4000	0.4683	0.5425	0.5897	18	0.4014	0.4716	0.5501
0.3077	0.3887	0.4555	0.5285	0.5751	19	0.3912	0.4596	0.5351
0.2992	0.3783	0.4438	0.5155	0.5614	20	0.3805	0.4466	0.5218
0.2914	0.3687	0.4329	0.5034	0.5487	21	0.3701	0.4364	0.5091
0.2841	0.3598	0.4227	0.4921	0.5368	22	0.3608	0.4252	0.4975
0.2774	0.3515	0.4133	0.4815	0.5256	23	0.3528	0.4160	0.4862
0.2711	0.3438	0.4044	0.4716	0.5151	24	0.3443	0.4070	0.4757
0.2653	0.3365	0.3961	0.4622	0.5052	25	0.3369	0.3977	0.4662
0.2598	0.3297	0.3882	0.4534	0.4958	26	0.3306	0.3901	0.4571
0.2546	0.3233	0.3809	0.4451	0.4869	27	0.3242	0.3828	0.4487
0.2497	0.3172	0.3739	0.4372	0.4785	28	0.3180	0.3755	0.4401
0.2451	0.3115	0.3673	0.4297	0.4705	29	0.3118	0.3685	0.4325
0.2407	0.3061	0.3610	0.4226	0.4629	30	0.3063	0.3624	0.4251
0.2070	0.2638	0.3120	0.3665	0.4026	40	0.2640	0.3128	0.3681
0.1843	0.2353	0.2787	0.3281	0.3610	50	0.2353	0.2791	0.3293
0.1678	0.2144	0.2542	0.2997	0.3301	60	0.2144	0.2545	0.3005
0.1550	0.1982	0.2352	0.2776	0.3060	70	0.1982	0.2354	0.2782
0.1448	0.1852	0.2199	0.2597	0.2864	80	0.1852	0.2201	0.2602
0.1364	0.1745	0.2072	0.2449	0.2702	90	0.1745	0.2074	0.2453
0.1292	0.1654	0.1966	0.2324	0.2565	100	0.1654	0.1967	0.2327

Notes

Notes

Notes

Notes